Global Perspectives on Health Geography

Series editor

Valorie Crooks, Department of Geography, Simon Fraser University, Burnaby, BC, Canada

Global Perspectives on Health Geography showcases cutting-edge health geography research that addresses pressing, contemporary aspects of the health-place interface. The bi-directional influence between health and place has been acknowledged for centuries, and understanding traditional and contemporary aspects of this connection is at the core of the discipline of health geography. Health geographers, for example, have: shown the complex ways in which places influence and directly impact our health; documented how and why we seek specific spaces to improve our wellbeing; and revealed how policies and practices across multiple scales affect health care delivery and receipt.

The series publishes a comprehensive portfolio of monographs and edited volumes that document the latest research in this important discipline. Proposals are accepted across a broad and ever-developing swath of topics as diverse as the discipline of health geography itself, including transnational health mobilities, experiential accounts of health and wellbeing, global-local health policies and practices, mHealth, environmental health (in)equity, theoretical approaches, and emerging spatial technologies as they relate to health and health services. Volumes in this series draw forth new methods, ways of thinking, and approaches to examining spatial and place-based aspects of health and health care across scales. They also weave together connections between health geography and other health and social science disciplines, and in doing so highlight the importance of spatial thinking.

Dr. Valorie Crooks (Simon Fraser University, crooks@sfu.ca) is the Series Editor of Global Perspectives on Health Geography. An author/editor questionnaire and book proposal form can be obtained from Publishing Editor Zachary Romano (zachary.romano@springer.com).

More information about this series at http://www.springer.com/series/15801

Alak Paul

HIV/AIDS in Bangladesh

Stigmatized People, Policy and Place

Alak Paul
Department of Geography and Environmental Studies
University of Chittagong
Chittagong, Bangladesh

ISSN 2522-8005 ISSN 2522-8013 (electronic)
Global Perspectives on Health Geography
ISBN 978-3-030-57652-3 ISBN 978-3-030-57650-9 (eBook)
https://doi.org/10.1007/978-3-030-57650-9

This Springer imprint is published by the registered company Springer Nature Switzerland AG
The registered company address is: Gewerbestrasse 11, 6330 Cham, Switzerland

To
Emeritus Professor Peter J Atkins,
Durham University, UK
(Dr Peter Atkins has been supporting
Bangladeshi Geographers as a true friend
for the last three decades)

Foreword

The Department of Geography at Durham University has welcomed many Bangladeshi research students over the last 50 years. Returning home most have taken up leadership positions in the university sector and their contributions to academic research and to applied research through consultancy have been notable.

Examples of important recent books by Durham alumni in Bangladesh include the monumental *National Atlas of Bangladesh* (Asiatic Society of Bangladesh 2017) jointly edited by Nurul Islam Nazem; Manzurul Hassan's *Arsenic in Groundwater: Poisoning and Risk Assessment* (CRC Press 2018); Nahid Rezwana's *Disasters, Gender and Access to Healthcare: Women in Coastal Bangladesh* (Routledge 2018); and *Geography in Bangladesh: Concepts, Methods and Applications* (Routledge 2019) edited by Sheikh Tawhidul Islam and Alak Paul.

Alak Paul is an expert in public health from the social and environmental perspectives with many publications in international journals. The present monograph is highly significant. I wrote in May 2020 at the peak of the pandemic of COVID-19, when all other news has disappeared from the media. It is timely that Alak's book reminds us that the HIV/AIDS pandemic has not gone away. It remains a threat despite falling from the headlines in recent years. His methodology of intensive qualitative fieldwork is impressive because it shows the human tragedies behind the infection. Reading the testimonies of some of his interviewees is harrowing and also moving, particularly because they were so surprised that anyone should take an interest in their plight. This type of work reminds us that medical epidemiology cannot give us a full picture of disease; we need to understand the social context of transmission in order to fashion appropriate policy responses.

Following Professor Paul's example will be a challenge for researchers who, these days, are required to produce quick results. Rather, this book is a reminder that fieldwork is often time-consuming and that the best scholarship requires commitment, determination and alertness to the human scale of the issues under consideration.

Em. Professor of Geography, University of Durham
Durham, UK

Peter J. Atkins

Acknowledgements

Though updated, this book is mainly based on the author's PhD thesis. The author extends his deepest gratitude to his PhD mentors (Professor Dr Peter J. Atkins and Dr Christine E. Dunn) in the Department of Geography, Durham University, UK who stimulated his interest in marginalized groups and guided his every step of the way and opened eyes to contemporary concepts and theories in health geography. Their unhesitating support, careful reading and comments in the early drafts, advice, and critique have enabled him to clarify his qualitative data, sharpen his ideas and develop the organization of his thesis. Special thanks are due to Professor Dr Jonathan Rigg of Bristol University and Dr. Marcus Power of Durham University, UK for their valuable comments and suggestions in combining theoretical aspects and 'real data'. The author also shows his best gratitude to Valorie Crooks, series editor of *Global Perspectives on Health Geography* of the Springer Nature and the entire editorial team for comments and guidance throughout this 'transformation'.

The author would like to show his thanks to Mary Shepherd (Johns Hopkins University), Dr. K. Sarker (India), Dr. Mahbubur Rahman (Japan), Dr. Matiur Rahman (ICDDR,B)) and Dr. Tapan K. Nath (Malaysia) for their assistance during the PhD research. He is grateful to Professor Dr. Md. Shahidul Islam of Dhaka University; Professor K. Maudood Elahi of Stamford University, Dhaka; Dr. Sk. Tawhidul Islam and Ms. Nandini Sanyal of Jahangirnagar University, Dhaka; Dr Dwaipayan Sikdar of Chittagong University; and Simon McConway, John and Pauline Owen of Durham for their every help both in Durham and Bangladesh. He would like to express his sincere thanks to the authority of Chittagong University (CU), Bangladesh for allowing him study leave for PhD and all members of the teaching staff of the Department of Geography, CU.

His special gratitude to the 'marginalized' community of Bangladesh particularly sex workers, drug users and people living with HIV (PLWH) who gave him enough time unexpectedly and shared their sufferings without hesitation, which actually helped him to make this book today. They not only allowed him to tell their unspoken life story for this research but also thought him a friend and well-wisher. The author is only hoping that those who read this book will understand their sufferings and hardships through the interpretations of their lives which he has come to understand. Here, his special thanks go to many NGOs, GOB officials and local people in Jessore and Khulna who acted as his local guides to make the 'bridge of friendship' with the respondents. His thanks go to Mr. Solzar Rahman, Mr. Nitish C. Mandol and Ms. Farzana Kabir for their involvement to facilitate his accommodation and research assistants. His thanks are also due to his research assistants (Md. Atiqur Rahman, Bishnu Mallick, Sanatan Kumar, Arjun Mondal, Indrojit Kumar, Anjan Das, Mamun Al Hasan, Syamol and Ashfak Mahmud) for the way they have worked with him in the course of this study.

He is very grateful to Overseas Research Students Awards Scheme (ORSAS), UK and Department of Geography in Durham University for financial support during the PhD research. He is also grateful to Geography Alumni Association, Ustinov Travel Award, Durham University; Sidney Perry Foundation, UK; and Charles Wallace Bangladesh Trust, UK for their financial support for PhD fieldwork and for completion of the research. Finally, the author extends his indebtedness to all his family members for inspirations, especially his life partner Sumana Podder and two boys, Anirudho and Arindom.

Overview

Acquired immune deficiency syndrome (AIDS) is one of the most complex health and socio-economic problems today, leading to many adverse impacts on individuals, communities and societies. It has become increasingly concentrated among marginalized populations in the developing countries like Bangladesh, a poor country in South Asia. Bangladesh is a predominantly Muslim country, where it might be thought the human immunodeficiency virus (HIV) is unlikely to be a problem because of traditional and conservative mores. But the results of different surveillance rounds have shown Bangladesh to be at risk of an HIV epidemic. There are many vulnerability factors such as geographical location, trans-border mobility, poverty, stigma and discrimination that favour the spread or transmission of HIV/AIDS. Most studies in Bangladesh on HIV are medical in approach and generally ignore the socio-economic, cultural or geographic linkages of HIV. Much research has been carried out on sexuality, sexually transmitted diseases (STDs), drug use and awareness related to HIV infection, but a few investigations have contributed to understanding the 'lifeworlds' of vulnerable people, and the stigma of marginalized communities. In addition, a few research projects have attempted to see the role of place and mobility in relation to HIV risk in Bangladesh. These research gaps have left planners poorly equipped to design and implement HIV prevention strategies. The present book seeks to bridge this gap in understanding health risk behaviour in relation to prejudice, place and policy by exploring the issues of vulnerable and marginal people's lives which put them at risk of infection and also their coping strategies and how these are played out. The aim is also to gain an understanding of the perceptions of local civil society, people and policy planners in explaining the vulnerability of people to HIV and proposing mitigation measures.

This book uses a qualitative approach as a methodological research strategy, recognizing that policy, people and places make a difference. In other words, the researcher has tried to explore the issue from a socio-geographic point of view along with health and policy planning in his field work in Jessore, Khulna and Dhaka. The location of both Jessore and Khulna has 'geographical value' as they have ports, brothels, opium dens, large transport terminals and slums. This research has been carried out on three social groups: First, the researcher worked with people vulnerable and marginalized to HIV infection (i.e. sex workers, drug users, people living with HIV (PLWH) and transport workers including Indian truckers). Second, he tal*ked* with local elites or people in civil society (i.e. journalists, NGO personnel and local government officials). Finally, he worked with key personnel and policy planners, i.e. high Bangladesh Government (GOB) and Non-Governmental Organization (NGO) officials in Dhaka. Regarding the sampling frame, besides the NGO beneficiaries, he managed this challenging work by developing contact with marginalized people through 'snowball' sampling. Despite its topic being a sensitive issue, this research did not negatively impinge on his respondents from any ethical or moral point of view. To fulfil the research objectives, the work is fully based on qualitative methods for data collection (i.e. in-depth interviews, focus group discussions, participant observation and naturalistic observation) and data analysis (i.e. grounded theory). The researcher used flexible conversational techniques for questioning the participants in convenient places.

The evolving HIV/AIDS pandemic has shown a consistent pattern through which marginalization, discrimination, stigmatization, and, more generally, a lack of respect for human rights and dignity of individuals and groups heighten their vulnerability. In particular, due to this social, economic and legal context, sex workers and drug users are subjected to harassment which can increase their 'everyday' vulnerability to sexually transmitted diseases and make them 'victims' to violence. Most PLWH participants felt a loss of self-image and self-esteem, uncertain and unpredictable future and distressing emotions. Discrimination against them has also been increasing. The qualitative information in this research demonstrates the real health risk to HIV/AIDS of the vulnerable people through their 'lifeworlds'. This research also managed to highlight or distinguish the geographically significant places, like port areas of Bangladesh in relation to STD/HIV/AIDS. It makes a relationship between geographic space and health risk, particularly with drug users and sex workers, through 'risk bridging'. Apart from women-trafficking and Indian truckers, this research has also found much evidence that many vulnerable people including sex workers, drug users and transport workers are at health risk due to their high mobility and the role of risky and non-risky places. A transparent and accountable mechanism is needed to ensure stronger coordination of activities on HIV and to ensure that commitments to HIV prevention and control are effectively translated into action. The government must formulate and implement programmes to reduce stigma and discrimination so that people living with HIV, and particularly members of vulnerable groups, can access services for prevention, care and support. It is expected that this research will ultimately lead to a better understanding of the social and geographical context of HIV/AIDS and provide a better foundation for health planning. In addition, this research develops a methodology of investigation for the study of a complex health and social environment in Bangladesh.

Regarding the arrangement of the book, a review of existing the HIV risk and vulnerability factors, descriptions of HIV literatures in Bangladesh and research gap, and detailed methods will be elaborated in the first chapter. Chapter 2 will look at the everyday geography of marginalized people's daily rituals, physical discomfort, sorrows and anger, dreams and frustrations which fuel their vulnerability.

Marginalized people's stigma, identity, rights and sufferings caused by discrimination will be explored in the third chapter. Chapter 4 will investigate the risk behaviours of vulnerable people and their level of knowledge about health risks along with different risk-minimizing techniques. Chapter 5 will examine the role of 'place' in channelling potential risk and the importance of geographical location in determining the hidden disaster of HIV in Bangladesh. Chapter 6 will discuss issues related to HIV prevention including addressing marginalization and stigmatization issues. A summary of the geographies of HIV/AIDS in Bangladesh will be discussed in Chap. 7 from the context of stigmatized people, policy and place.

Abbreviations

AIDS Acquired immune deficiency syndrome
ART Antiretroviral therapy
BDR Bangladesh Rifles
BSF Border security force
BSS Behavioural surveillance survey
CSW Commercial sex worker
DFID Department for International Development
DU Drug user
FGD Focus group discussion
GOB Government of Bangladesh
HAPP HIV/AIDS Prevention Project
HBV Hepatitis B virus
HCV Hepatitis C virus
HIV Human immunodeficiency virus
HNPSP Health Nutrition Population Sector Programme
ICDDR, B International Centre for Diarrhoeal Disease Research, Bangladesh
IDH Infectious Diseases Hospital
IVDU Intravenous drug user
KABP Knowledge, attitudes, beliefs and practices
MR Menstrual regulation
MSM Males sex with males
NASP National AIDS/STD Programme
NGO Non-Governmental Organization
OI Opportunistic infection
PBD Professional blood donor
PLWH People living with HIV/AIDS
STDSexually transmitted disease
STISexually transmitted infection
Taka1 taka= 0.012 US $
UBINIG Unnayan Bikalper Nitinirdharoni Gobeshona
UNAIDSJoint United Nations Programme on HIV/AIDS

UNFPA	United Nations Population Fund
USAID	United States Agency for International Development
VCT	Voluntary counselling and testing
WHO	World Health Organization

Contents

Chapter 1
HIV/AIDS in Bangladesh and Present Research

1.1 Introduction

HIV/AIDS is one of the most complex health and socio-economic problems in the world at present, having adverse impacts on individuals, communities and societies. Over the last two decades, it has become increasingly concentrated among marginalized populations in developing countries like Bangladesh. Apart from behavioural and bio-medical risk factors, HIV/AIDS has spread fast where there is widespread stigma and discrimination, along with poverty and illiteracy. In particular, stigma continues to remain a major barrier to treatment and this in turn enhances vulnerability. According to the World Health Organization (2019), HIV is transmitted mostly through semen and vaginal fluids during unprotected sex without the use of condoms. Globally, most cases of sexual transmission involve men and women, although in some developed countries homosexual activity remains the primary mode. Beside sexual intercourse, HIV can also be transmitted during drug injection by the sharing of needles contaminated with infected blood, by the transfusion of infected blood or blood products and from an infected woman to her baby: before birth, during birth or just after delivery (WHO, 2019). Many people with HIV do not know that they are infected and this lack of diagnosis makes it difficult to bring them under any form of care. Once infected, a person may not have symptoms for many years but can still transmit the disease to others. The virus multiplies in the body and eventually destroys the immune system. As a result, tuberculosis and other bacteria can cause opportunistic infections (OIs), although usually these organisms will not cause disease in healthy people. The terminal stage of HIV infection, when patients suffer from OIs, is called AIDS. Approximately 50% of HIV-infected persons will develop AIDS after 7–10 years of infection. The average survival time for a person with AIDS may be only 6 months in developing countries and 1–3 years in developed countries (WHO, 1997). However, with the advent of new antiretroviral therapy, survival has improved dramatically in richer countries. These drugs, which are

A. Paul, *HIV/AIDS in Bangladesh*, Global Perspectives on Health Geography,
https://doi.org/10.1007/978-3-030-57650-9_1

very expensive, are unfortunately beyond the reach of most patients in the developing world. According to the World Disasters Report by International Federation of Red Cross (IFRC, 2018), HIV/AIDS is the disaster that keeps on killing.

Globally, the HIV/AIDS epidemic constitutes one of the most burning threats known to humankind. Since the start of the epidemic, more than 35.4 million people have died from AIDS-related illnesses and 77.3 million people have become infected with HIV. According to the latest figures published by UNAIDS (2018), an estimated 37.9 million people globally were living with HIV. 1.7 million people became newly infected with HIV in 2018 and about 1 million people have died from AIDS-related illnesses in 2018. About 23.3 million people were accessing antiretroviral therapy in 2018. More than three quarters of AIDS-related deaths occur in Sub-Saharan Africa, and South Africa is the country with the highest prevalence of HIV in the world. Currently, 66% of the total people with HIV infection are concentrated in Sub-Saharan Africa, but epidemics elsewhere in the world are growing rapidly. More than 90% of the PLWH population resides in developing countries (WHO, 2019). Epidemiological studies indicate that, unlike the western world and Africa, HIV is a relative newcomer to Asia. In South East Asia, it was first identified in 1984 in Thailand. In India, it was first reported in 1986, in Burma in 1987 and in Bangladesh in 1989 (Paul, 2009). It is estimated that the region had 3.8 million HIV-infected persons in 2018 (UNAIDS, 2018). Exploratory sero-surveillance indicates an epidemic scale of HIV infection in Thailand, India and Burma, especially in population groups engaged in unprotected heterosexual sexual activities. Traditionally, CSWs in South Asia have been brothel-based. But, due to rapid socio-economic and cultural change, the commercial sex business has undergone significant change. CSWs are now available in hotels, restaurants, bars, street corners or inconspicuous houses in residential areas, massage and beauty parlours. In addition, intravenous drug use is high in Burma, Nepal and north east India, all of which are very close neighbours of Bangladesh. In South Asia, the HIV situation in India, Nepal and Burma is critical, in terms of the number of HIV patients and risk. With many people living with HIV/AIDS in neighbouring countries, what is the situation of a Muslim nation, Bangladesh? Although it is a predominantly a culturally conservative Muslim country, there are many social and economic, as well as geographical factors, fuelling the potential HIV risk in Bangladesh through insecure sexual activity and drug habits which in theory are 'visibly' prohibited.

1.2 Risk and Vulnerability Factors for HIV/AIDS in Bangladesh

Being a third world country, Bangladesh is highly vulnerable to HIV/AIDS, not only as a medical challenge, but also in terms of its socio-cultural and geographical context (Paul, 2009). Bangladesh is one of the four countries in the region where the epidemic continues to increase (UNAIDS, 2012). Though Bangladesh has a low

HIV prevalence (less than 1%) in the general population (NASP, 2014), according to the National Surveillance of 2015–2016, a concentrated epidemic has been recorded among the male drug users in a neighbourhood of Dhaka (old Dhaka) where the prevalence was 27.3%. The prevalence of HIV among sex workers and other risk groups is less than 1% (NASP, 2016). The first case of HIV in Bangladesh was detected in 1989 and up until December 2016 the total number of detected cases was 4721 of whom 799 have died, leaving 3922 known people living with HIV (NASP, 2016). About one third of detected PLWH are women. Among the 578 new HIV cases reported in 2016, 32.9% were among women, and 5.5% were among children. However, the majority of infections are likely to remain undetected, and the total national estimate is about 14,000 PLWH (UNAIDS, 2018).

The country has a range of contextual factors that create and sustain this vulnerability. A stereotypical view of the people of Bangladesh is that they are likely to follow Islamic religious norms meticulously as it is a predominantly Muslim nation. There is a common misconception, for instance, that STDs, HIV and AIDS are not health risks for them. But in reality there is evidence that all of the health risk behaviours related to HIV (e.g. premarital/extra marital sex, homosexuality, prostitution and intravenous drug use) are present, contradictory to the norms of the mainstream, conservative society (see Khan, Hudson-Rodd, Saggers, & Bhuiya, 2005). In addition, there are also many vulnerability factors (e.g. geographical location, trans-border mobility, low HIV/AIDS awareness, poverty and gender inequalities), which individually and collectively favour the spread or transmission of HIV/AIDS (see Gibney et al., 1999). Although Bangladesh is considered a low prevalence nation at present, from the epidemiological point of view, the HIV situation is evolving rapidly. Importantly, it is commonly assumed that some significant geographical locations, particularly the urban and border areas are channels for the 'transmission' and 'importation' of HIV into the country where CSWs, IDUs and TWs are highly mobile and sell sex and share needles. As a consequence, there are several key nodes for the diffusion of the disease across the whole country. In what follows, the existing risks and vulnerabilities, including behavioural, bio-medical, social and structural, for HIV infection in Bangladesh are explained.

Sexuality AIDS is overwhelmingly a sexually transmitted disease and sex is surrounded by taboos in nearly all human societies, including Bangladesh. Many research findings mainly by medical scientists and sociologists indicate that the incidence of extramarital sex is quite widespread in Bangladesh (Rob & Mutahara, 2000). Prostitution is prohibited, but there are a significant number of female sex workers, particularly in the urban, border and port areas. Studies have shown that Bangladeshi society, long considered so conservative, is more footloose and sexually free than is commonly admitted. Aziz and Maloney (1985) found that 50% of youths, mostly of the lower socio-economic class, have experienced sex before marriage. Folmer et al. (1992) also found prevalence of premarital sex among their respondents with 29%of them using condoms. Other researchers such as Islam (1981) and Maloney et al. (1981) reported similar findings in the significant proportion of their subjects and noted occurrences of induced abortions among unmarried girls. According to the World Bank Report (2006), a flourishing commercial sex industry is an important behavioural risk for HIV in Bangladesh. A few important academic works on behavioural risks, by Caldwell and Pieris (1999) and Roy, Anderson, Evans, and Rahman (2010) found evidence which explained many anomalies regarding social issues on sexuality and also evaluated the prospects of an AIDS epidemic. They found that levels of premarital and extramarital sex among men, especially transport workers, are moderate by international standards but probably higher than the expectation in a socially conservative society. However, about 8–11% of Bangladeshi men aged 15–49 years (NASP, 2009) buy sex from female sex workers, occasionally or frequently. Apart from lower class men, recently it is an open secret that many of the middle-high and high income people frequently engage in hotel/residence-based extramarital sex in the metropolitan cities (Gazi et al., 2009).

Deteriorating economic conditions in Bangladesh are leading more and more Bangladeshi women to the sex trade. They are generally non-literate, divorced or separated women and may be organized in brothels or may be ‘floating’ (Khan, 1999). Floating sex workers waiting to be picked up is a common nightly scene in many areas of big cities, including Dhaka and Chittagong. By comparison, MSM behaviour is largely hidden; it is, however, believed that it is more prevalent than previously thought (Chan & Khan, 2007; Khan et al., 2005). There is some evidence of homosexuality among labourers, transport workers and boys. Particularly, rickshaw pullers and manual cart pullers have been reported as committing rape on a certain group of child labourers who sleep in an open part of the market at night (Choudhury, Arjumand, & Piwoo, 1997).

Illegal Opiate Use Drug trafficking is the distribution of illicit drugs by large-scale operations, which can, and often do, cross national boundaries, as well as the small-scale syndicates that distribute drugs at the local level (Bean, 2002). Afghanistan accounts for almost 75% of the world’s illicit opium supply (MacDonald & Mansfield, 2001). Much of the remainder is from the traditional growing region of the Golden Triangle (Burma, Laos and Thailand). Significant amounts, however, are grown elsewhere, such as in Iran and Turkey. Bangladesh is considered to be an important hub of illicit drug smuggling. It is located between the ‘Golden Triangle’ and the ‘Golden Crescent’ opiate-producing zones, and has become an easily accessible market for opiates (GOB, 2002). The problem of drug abuse has reached recognizably significant proportions today in Bangladesh and it is linked to organized and petty crime (Muntasir, 2005). In general, most addicts are males. Frustration, curiosity and peer pressure are the most frequent reasons given for drug addiction. Illegal opiate use behaviour, which is considered a lifestyle risk factor for HIV, is prevalent. In addition, a significant number of drug users are extremely marginalized and live on the streets and out of any social structure which puts them in more vulnerable situations. Repeated rounds of surveillance have revealed that the rate of sero-positivity is highest among intravenous drug users (IVDUs) and the findings also confirm the presence of high levels of behavioural risk factors for the acquisition of HIV infection through needle sharing (Azim et al., 2008; Mondal, Takaku,

Ohkusa, Sugawara, & Okabe, 2009). Injecting drug use has steadily gained in popularity in Bangladesh (Khan, 2006). A considerable proportion of heroin users shared a needle/syringe during their last injection. Recent Behavioural Surveillance Survey (BSS) data indicate that the drug user population is well integrated into the surrounding urban community, socially and sexually, thus raising concerns about the spread of HIV infection (Islam et al., 2015). However, drug users are also sexually active with their married or unmarried partners. BSS data also indicate an increase in risk behaviours such as sharing of injecting equipment and a decline in consistent condom use in sexual encounters between drug users (including IVDUs and heroin smokers) and female sex workers. More than half of the heroin users had commercial and non-commercial female sex partners in the last year and those who did had multiple partners. Condom use, both in the last sex act and consistently in the last month, was very low with both commercial and non-commercial partners. This overlapping, as well as multiple relations between more vulnerable and bridging populations, makes Bangladesh vulnerable to HIV/AIDS.

Awareness of HIV/AIDS Public awareness of HIV/AIDS is an important prerequisite for behavioural change. Many people in Bangladesh have now heard the words 'HIV' and 'AIDS' as a result of intensive campaigns by different media and NGOs (Rahman & Rahman, 2007), but a significant portion still need to know how it is transmitted and that it can be prevented. The level of awareness among high-risk groups, as well as the public generally, is surprisingly low (Yaya, Bishwajit, Danhoundo, Shah, & Ekholuenetale, 2016). Educated people are better informed about HIV prevention than the poor but, as Caldwell and Pieris (1999) explain, the lack of openness with regard to sexual issues means that opportunities for improved knowledge are limited. Regarding awareness of STDs among married women of reproductive age, Khan, Rahman, Khanam, Barkat-e-Khuda, and Ashraf (1997) found that women in rural areas are generally not aware of STDs, although their knowledge varied with their husband's occupation, and their own age and level of education. But Haque, Hossain, Chowdhury, and Uddin (2018) mentioned that a sizeable portion of married women have an adequate knowledge and awareness regarding the HIV/AIDS. Bhuiya, Hanifi, Hossain, and Aziz (2000) showed that awareness of HIV/AIDS among the general population, particularly villagers, is very low. The total number of non-marital sexual contacts as well as the low knowledge levels/exposure to HIV prevention information and programming suggests that community-wide awareness campaigns are needed. Epidemiological data, particularly on STIs and related complications among high-risk behaviour groups, is limited in Bangladesh due to poor recognition of STIs as a major health problem, the stigma and discrimination associated with STIs and the lack of coordination between service providers and the research community (Nessa et al., 2005). However, different surveys of long-distance truck drivers revealed that most of them had contact with commercial sex workers (CSWs) about twice a month but that they mostly do not know about HIV/AIDS.

Blood Transfusion System Unscreened blood transfusions and commercial blood donation by professional donors are important bio-medical risk factors for HIV infection in Bangladesh. Many studies shown that blood transfusion could be a potential risk in future as blood screening facilities are not good in most hospitals and blood banks, and there are no effective rules for commercial blood collection, testing, processing, storage and distribution (i.e. Akhter et al., 2016; Hossain, Bhuiya, & Streatfield, 1996). The existing blood transfusion system carries a danger of HIV transmission. Currently, the country needs annually about 300,000–350,000 units of blood, only 20–25% blood comes from voluntary donation. A significant percentage of (about 30%) blood is collected from professional blood donors through illegal blood banks (The Independent, 2017). Most professional blood donors (PBDs) sell their blood to raise money to feed their drug habits, which can also fuel the danger of HIV transmission. It is suspected that a good number of PBDs are positive for hepatitis B and syphilis (Alam, Hossain, & Chowdhury, 2015). Recently, some PBDs have been identified as PLWH in a border area of the country. Moreover, unsterile injections in non-formal and formal health-care settings can be one of the significant potential bio-medical risk factors for HIV infection in Bangladesh (Gibney, Choudhury, Khawaja, Sarker, & Vermund, 1999). It is a fact that disposable plastic syringes including needles are resold in shops and markets, adding greatly to the risk of spread of HIV/AIDS.

Sexually Transmitted Diseases A high prevalence of sexually transmitted diseases (STDs) is a co-factor for HIV transmission among the 'most at-risk' groups, and is associated with bridging to the general public if untreated or ineffectively treated. Bangladesh has a large commercial 'sex industry' in different forms. A diverse group of clients buy sex and only a few use condoms. Many believe that the low level of HIV/AIDS in the Muslim dominated society is a result of societal conservatism and religious teaching about immoral sex. But there are studies showing high levels of commercial sex and significant levels of STDs, thus presenting a situation more complex than the stereotype (Rahman & Rahman, 2007). Research confirms that there is a low level of condom use by sex workers (Sarkar, Islam, Durandin, et al., 1998). These CSWs are mostly non-literate and are organized in brothels or work casually in the streets (Khan, 1999) but condom use for commercial sex is low in Asia (only 2%) and the turnover of clients is high (averaging 44 clients a week). As a result, most CSWs have STDs, including syphilis and gonorrhoea (Rahman, Alam, Nessa, Hossain, et al., 2000; Nessa et al., 2005). A number of studies (Hossain, Mani, Sidik, Shahar, & Islam, 2014; Khanam et al., 2017; Sabin, Hawkes, Rahman, et al., 1997) also indicate the prevalence of STDs and RTI over a long period in the country. In addition, high rates of STDs also appear to be present among drug users. The existence of risky behaviour and high levels of sexually transmitted infections (STIs) among the 'core groups' indicates the potential for a serious HIV/AIDS epidemic in Bangladesh. However, transport workers are not necessarily regarded as a vulnerable group in terms of national surveillance. They are considered to be possible 'bridges' to the general population, potentially transmitting HIV and STIs to their wives and to the population in general (Knight, 2006).

Social and Structural Risk Bangladesh has many contextual features, including widespread poverty, gender inequality, stigma and discrimination, violence, poor healthcare infrastructure, untrained health care personnel and low levels of literacy that are relevant to HIV risk and vulnerabilities. Poverty and gender inequalities have been playing an important role in transmitting the risk among marginalized people. Poverty is the primary cause of trafficking in the region and traffickers target their prey in the poverty-stricken rural areas. Due to poverty mainly, human trafficking into prostitution, stigmatization of these women, conservative social attitudes and huge migration flows (mainly rural to urban) exist (Mahmood, 2007). The HIV/AIDS epidemic in Asian countries has been strongly influenced by gender inequality and the frequent practice of men visiting sex workers. Women lack the power to refuse sexual activities due to a lack of economic empowerment and the cultural convention that wives are unable to refuse sex with their husband or demand the use of a condom.

The biggest challenge to an expanded response to HIV/AIDS in Bangladesh is the government's limited funding capacity. There are limited care and support provisions for PLWH (NASP, 2004). Regarding health care, many health care personnel do not have appropriate training to handle the medical needs of people living with HIV/AIDS. Social and cultural barriers in risk prevention are formidable. Stigma and discrimination are problems, and public perceptions of PLWH and members of vulnerable populations are negative (Panos, 2006). However, HIV/AIDS policy should not only emphasize medical and technological aspects, but should also be based on social and economic considerations. Human rights must be addressed in a comprehensive national policy. In all international declarations and the national policy on HIV/AIDS, reference is made to the need for a human rights framework. Human rights here include access to health care, information, confidentiality and gender equity (NASP, 2005).

Geographical Location and Mobility The geographical nature of Bangladesh, in particular its long borders with India and Myanmar, exacerbates the HIV/AIDS hazard (Gibney, Choudhury, Khawaja, Sarker, Islam, & Vermund, 1999; Sarkar et al., 2008). Bangladesh has a significant cross border trade at land ports and movements of population, including high-risk groups between Bangladesh and India. This trans-border mobility is high for various reasons, i.e., trade, education, religious exchange, recreation and some illegal activities. It is notable that India is in the phase of rapidly rising prevalence of HIV, with an estimated 2.1 million cases (UNAIDS, 2017). Bangladesh has thirty border districts, 28 sharing a border with India and two with Myanmar. Most of the land frontiers are open with rivers running across. Bangladeshi trafficking groups have been able to build up powerful bases in the border districts of India in West Bengal and Assam, to the north and west, and these are now the favourite transit points for trafficked people (GOB, 2004; IOM, 2004). Population increases, environmental crises and structural adjustments in Bangladesh have encouraged migration to India. An estimated 2000 Bangladeshis cross the border every day, including labourers, smugglers and trafficked women and girls (Knight, 2006). Porous borders with economically poorer Bangladeshis (not needing a visa to visit India) aggravate the problem of cross border trafficking and the country has remained a source of women and children for some considerable time. The environmental closeness of Bangladesh to Myanmar, and consequently the Golden Triangle drug trail, has made it a major transit route for drug smuggling. The HIV epidemic among drug users in Myanmar and its heroin export routes has led to HIV epidemics in neighbouring countries (Chelala & Beyrer, 1999). Myanmar has been considered a primary contributor to the spread of HIV/AIDS in this region and 240,000 cases were estimated of HIV infection in Myanmar by the end of 2018 (UNAIDS, 2018). However, the constant movement is one of the major reasons for the transmission of HIV particularly among drug users in northern India, Afghanistan, Pakistan and Bangladesh. In addition, there is a high prevalence of HIV cases in two of India's north eastern provinces, Nagaland and Manipur, which have Myanmar as a neighbour. China is also facing a similar crisis along the stretches where it shares Myanmar's north eastern border (UNAIDS, 2006).

However, Bangladesh has a large number of overseas migrant workers who have gone in search of better job opportunities mainly to countries in the Middle East, or Malaysia and South East Asian countries (Mercer, Khanam, Gurley, & Azim, 2007). It is widely suspected that some of them have come back after being infected with STDs and HIV from these countries. Most of the detected HIV/AIDS cases in Bangladesh are overseas migrant workers. In addition, there is a great deal of migration between rural and urban areas within the country (NASP, 2004), and Bangladesh hosts large communities of *Rohingya* refugees from Myanmar in the south east part of the country. The two major seaports also receive many foreign ships' crews the year round. Together, these population movements add to the risk of STDs and HIV/AIDS.

1.3 Research on HIV/AIDS in Bangladesh and Present Endeavour

In much research, HIV/AIDS continues to be considered as a 'bio-medical/behavioural level' problem, but Campbell and Williams (1999) suggest that it should be seen instead as a 'Bio-Psycho-Social' problem. In other words, epidemiological data alone cannot represent its multiple and complex social dimensions (Mann, 1987). The critical interacting factors of sexual and injecting drug-using cultures, together with the governmental and societal response to the HIV threat, shape the geographies of HIV/AIDS in particular settings (Ford, Siregar, Ngatimin, & Maidin, 1997). The HIV epidemic has brought new dimensions of risk to third world countries like Bangladesh. It has also brought many changes, not only in the health sector, but also in social and economic issues (NASP, 2016). These changes have occurred in the everyday lives of the individuals and communities who are highly vulnerable or affected by the HIV threat. Over the last 25 years, around the world many geographers, sociologists, anthropologists, psychologists, epidemiologists and medical scientists have had an increasing interest in HIV and AIDS and its related subjects. In Bangladesh, this influence has been very limited. Most of the research on HIV has been performed by public health scientists and epidemiologists. Very little academic research on HIV has addressed the socio-economic or socio-geographic issues of the marginalized and stigmatized communities who are considered as the 'risk group' for HIV infection in Bangladesh.

There is a vast literature on epidemiology and aetiology for the most at-risk groups. These reports have looked not only at HIV infection but also at STDs and hepatitis. Firstly, most work (for example, Azim et al., 2000; Bosu, 2013; Hossain, Akter, Kamal, Mandal, & Aktharuzzaman, 2012; Islam, Hossain, Kamal, & Ahsan, 2003; Mowla & Sattar, 2016; Nessa et al., 2005) has assessed risk through blood testing. Secondly, much research (for example, Alam et al., 2013; Gibney, Choudhury, Khawaja, Sarker, Islam, & Vermund, 1999; Haseen et al., 2012; Jenkins, 1999; Kamal, Hassan, & Salikon, 2015; Rahman & Zaman, 2005) has assessed risk behaviours for HIV prevalence of sex workers and drug users as well as truckers. A few works (like, Gibney, Choudhury, Khawaja, Sarker, & Vermund, 1999; Mollah et al., 2004) have also looked at the bio-medical issues of HIV risk in Bangladesh. These above-mentioned studies have tried to show the potentiality of the future HIV threat to Bangladesh from the behavioural as well as the bio-medical point of view. Thirdly, there are a good number of papers (for example, Islam, Mostafa, Bhuiya, Hawkes, & de Francisco, 2002; Khan et al., 1997; Sarafian, 2012) which have focused on the assessment of awareness of HIV/AIDS among the different stakeholders, especially the most at-risk groups for HIV. Finally, all the above categories of work discuss some relevant preventive efforts in their concluding remarks.

With a dominant bio-medical and epidemiological framework, many studies of HIV in Bangladesh frequently ignore the explanatory issues regarding the prejudice against marginalized and vulnerable people, unhealthy places and policy, all of which can be investigated by utilizing qualitative methodologies (Paul, 2019). Following Mann (1987), epidemiological data alone cannot represent the multiple and complex social dimensions of HIV/AIDS. According to him, the pandemic has been conceptualized as consisting of three separate phases: an epidemic of HIV infection, an epidemic of AIDS and an epidemic of social, cultural, economic and political responses to AIDS. The third of these has been the most explosive, characterized by denial, stigma and discrimination. However, a substantial literature supports the notion that the HIV/AIDS epidemic is determined by a combination of structural, social/cultural and individual factors (Mac-Phail & Campbell, 2001; Scambler & Paoli, 2008). The AIDS pandemic is now extremely complex, consisting of a number of smaller and constantly changing epidemics which affect individuals, communities and nations in a multiplicity of different ways. But in Bangladesh, many arenas of research in the HIV field have not been covered equally. It has been found in this review that there has not yet been any in-depth research concerning the socio-cultural and geographic impacts of the HIV disaster in Bangladesh. Almost all of the literature shows HIV as an epidemiological problem rather than investigating it from a social or cultural point of view and still less using qualitative methods. The present research is an endeavour to fill these gaps. The valuable qualitative field data will demonstrate the causes of HIV risk and vulnerability, and the book seeks a better understanding of the nature of the social and locational context of HIV/AIDS in Bangladesh and will therefore assist with health care policy planning. Although most epidemiological research (for example, Gazi, Mercer, et al., 2008; Hosain & Chatterjee, 2005) has hinted at the necessity for doing intensive study on geographically significant places like the border towns of Bangladesh, there has not been significant academic research in this area. Furthermore, there has been no work on PLWH and their 'lifeworlds' in conjunction with HIV policy issues in Bangladesh. Little academic research (see, for example, Hasan, 2007; Hossain & Kippax, 2011; Human Rights Watch Report, 2003; Sarma & Oliveras, 2011; Paul, Atkins, & Dunn, 2012; Karim & Mona, 2013) has been carried out from the social, geographical or policy point of view.

The aim of this study is to explore those issues in vulnerable and marginal people's lives which put them at risk of infection and also their adopted coping strategies and how are these played out. In addition, the research aims to gain an understanding of the perceptions of civil society and policy planners with respect to vulnerability to HIV and the necessary mitigation measures. Within these general aims, the study has five main objectives. First, I will detail the 'lifeworlds' of marginalized communities in terms of their everyday practices or customs, along with their emotions and aspirations. Here, I will consider brothel and non-brothel sex workers, opiate users and PLWH as marginalized groups because of their negative status, the hostility of mainstream society and active discrimination by the state. Their monetary uncertainty and everyday suffering are covered in detail. Their anger and expectations from society are also discussed which combined play a role in putting them at health risk. Conducting this research required an extended period of time to be spent with the marginalized communities to obtain the views about 'lifeworlds' through in-depth interviewing, focus group discussion and naturalistic observation. Second, there is no information on how HIV-affected or HIV-prone people in Bangladesh live with social hazards. Still less research in Bangladesh has focused on the way in which these social and psychological factors affect people's everyday lives.

This book is devoted to the task of studying the socio-geographical and psychological aspects of having HIV or living in close proximity to people with HIV. The investigation of identity 'crises' and related consequences due to marginalization and stigmatization status will be helpful in measuring their potential risk of infection. Victimization processes will be explored in order to understand the negative societal attitudes in each study area. Third, in this book, 'risk' is discussed in the light of the implications for understanding how everyday norms influence the ways in which people perceive risk and act in response to risk. Vulnerable people's perceptions about their risk behaviours and their impact on health will be of great help in revealing the real causes of risky behaviour in relation to HIV infection. Vulnerable people include sex workers, drug users and Bangladeshi and Indian truckers. In addition, their knowledge or consciousness about health risks, as well as their coping and adaptive techniques, will uncover their experiences of living with health risks. Fourth, this book will examine places from different points of view, such as those that are thought to be safe, risky or stigmatized. Through interviewing border girls and foreign truckers, I will cover the role of mobility of vulnerable people and the role of border towns in transferring the potential risk. Moreover, I will also look at the issue of probability of HIV risk for Bangladesh through trafficking and other unidentified sources. Fifth, a number of publications discuss different HIV prevention measures of Bangladesh but there is no information on the 'contradictory programmes' for HIV prevention in Bangladesh, 'standard awareness campaigns', stigma and discrimination for PLWH, NGO politics and donor policy, and the role of the government. Very little detailed research has explored HIV policy, policy implementation or the components of HIV prevention strategies in Bangladesh. This will help to identify the inherent policy weaknesses and assist in developing strong recommendations for preventing and mitigating this upcoming health disaster.

The countries of the global South that are facing major HIV/AIDS problems, including Bangladesh, have varied forms of health risk depending upon their contingent circumstances. I have found that writing this book has helped me to learn about the generic aspects of health risk, coping strategies, health facilities and health care for HIV/AIDS. I have also read widely about methodological and theoretical aspects and this has helped to plan the various dimensions of my research. But, practically speaking, I am aware that the situation on the ground in Jessore and Khulna is very different from the papers I have read about Africa or North America. Therefore, I built a framework for my work that is both aware of this HIV/AIDS work elsewhere but that is also geared towards the particular cultural, social and economic circumstances of my field area. It is mentionable that my research is totally based on 'lifeworlds' of marginalized people—those who are vulnerable to HIV. Stigma and rights issues are looked at in this study as well as dimensions of risk and place, rather than adopting a medical perspective. The research also pays attention to relevant policy issues related to HIV. Considering the preceding methodological and conceptual discussion, I have adopted qualitative methods for data collection and analysis.

1.4 Ethical Approach of the Research

As HIV/AIDS is highly stigmatized, a question may arise about the ethical issues of this kind of research. This work was about knowing the 'everyday lives' of participants and I was very careful to make sure that my research did not negatively impact on these respondents in any way. After approaching them, once I was confident that they were relaxed I asked for their patience and permission for recording. In the meantime, I assured them about the confidentiality of their conversation. At the beginning of each research encounter, I told my all respondents that they may face some shyness in replying to some 'sensitive' questions but I assured them that I would not ask their name and address. Particularly, some hotel or floating girls, or PLWH had some worries as they are local residents. Sometimes they asked me "*what is your profit*"? or "*Why will you keep it confidential*?" Then I described more detail about my research objectives and anonymity issues. Although I had given my reply, one of them asked me, "*There are many issues, so why are you doing your research on this critical topic*"? Then I replied that it is an important issue for the country in the near future. I managed to get some initially quiet participants to be very vocal in the interviews or FGDs when I gave the assurance of confidentiality. After getting this assurance, many participants were willing to divulge details of their 'lifeworlds', including their risk and vulnerability. But there is an exception with one PLWH. I thought that she felt too shy to speak anything about her HIV status and was trying to avoid my questions. Despite my assurance of confidentiality, she refused to give me an interview, giving the excuse that she had a headache. Then I stopped taping the interview and tried informal conversation instead and then she felt comfortable. But recently, I have heard that she has died. As a final question in all interviews and FGD cases, I asked my participants whether my interview topic or any of my conversation has given them any distress.

In this book, I have changed all respondents' names in the text. I often provided a complimentary gift or food or money as a form of compensation for participants' time, patience and cooperation. For example, in most of the cases of sex workers and female slum dwellers, I gave some gifts (such as body lotion, shampoo, soap or beauty cream) and offered food (ranging from light snacks to lunch). I gave some token money to the street girls who are very poor. For PLWH, I gave energy drinks (e.g. Horlicks) and fruit, along with some token money for transport costs since I interviewed some of them in NGO offices. All of them were very happy with this. I also offered some light snacks for the transport workers. In the case of drug users, I used different approaches. To some I gave money to help them buy food or sometimes I took them into a restaurant to offer lunch. Some begged money from me and many addicts broke down in tears when they took food or money. Though I gave them money for buying food, some may have used it for buying heroin. In one drug users' treatment centre, some of my FGD participants requested me to ask the officials to improve their food. I always provided token amounts of money for the transport costs to NGO peers who brought or managed the participants. I also offered lunches and dinners for NGO officials in Jessore and Khulna for their help. In many cases, I took photographs (by myself and research assistants) during my research but only after obtaining participants' permission beforehand. Many addicts, TWs and some street girls were keen to have their photos taken with me. In some cases, I gave them a photograph of themselves. But I never asked PLWH for photos because of their high stigmatization. In Khulna, I faced a difficult issue with one street girl who had recently found herself pregnant and who begged some money from me. Because of my own concerns that she may have used the money to help pay for an abortion, I decided not to offer money.

1.5 Approach to Study Area Selection

My reasons for selecting the study sites are mainly based upon 'purposive or theoretical sampling criteria' rather than a 'statistical probability approach'. The selection of study sites was primarily based on the conceptual framework of the HIV issues supported by the literature reviews, including a few published (mainly epidemiological) and unpublished (mainly newspaper clippings) materials. It is usually assumed that some important geographical locations, particularly the urban and border areas are channels for the 'transmission' and 'importation' of HIV, respectively, within the country and from neighbouring countries. The study areas (Fig. 1.1) are Jessore and Khulna, where many vulnerable people like CSWs and DUs are visible on the streets and many foreign and native truckers are present. Both places are therefore nodes of disease diffusion for the whole country.

I worked firstly in Jessore, a frontier district, and found a few brothels and hotels where many girls have been involved in the commercial sex profession. I also found many street drug users in Jessore town. In Benapole, frontier town under Jessore district, I worked to find transport workers particularly Indians as well as Bangladeshi women who are used to crossing the border almost every day. In Benapole, more than 300 Indian trucks approach every day and more than 500 Bangladeshi trucks arrive from different districts. Geographically, Benapole is a strategic point for border trade between India and Bangladesh and it is the most important check post and busiest land port on the Indo-Bangladesh border. Benapole land port is also lucrative for Indian exporters because of its cheap service and equipment charges. The second study area is Khulna, which is also geographically significant as an industrial and metropolitan city along with its port setting. I also found many identified PLWH in Khulna and this also persuaded me to select the place as my second study site. In Khulna, I went to Baniashanta and Fultola, the two local brothels. Finally, I also worked in Dhaka, the capital, to talk with policy planners on the HIV situation in Bangladesh.

1.6 Methods

Although a significant amount of empirical research on health and place has applied extensive, quantitative methods and techniques for statistical modelling of a general 'contextual' effect on the health of populations (Andrews & Evans, 2008; Dunn, 2007; Mason, 2007; Sui, 2007; Townley, Kloos, & Wright, 2009; Yiannakoulias, Svenson, & Schopflocher, 2005), some developments and changes of emphases have been accompanied and facilitated by methodological shifts from quantitative to qualitative which, in Kearns' (1995) emphasis, is 'recasting of the subjects of research as persons rather than as patients' (p. 252), using participant observation, in-depth interviews, focus groups, storytelling and autobiography as methods (Kearns, 1995). Wilton (1999) stated that 'in health geography, qualitative research

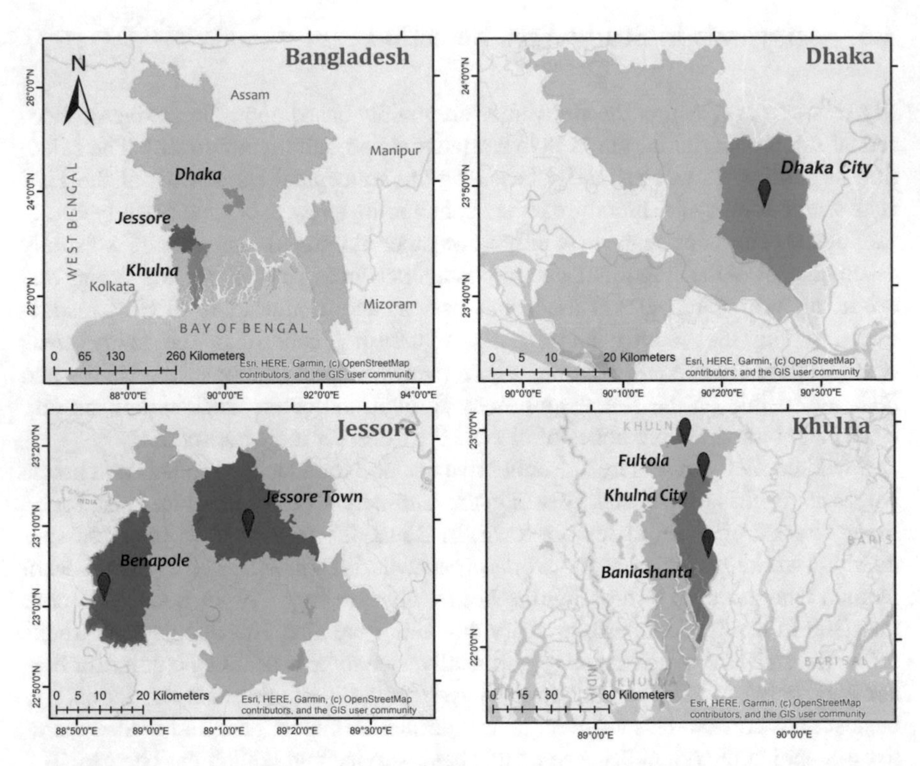

Fig. 1.1 Study area. The map is prepared using ArcGIS 10.5 and the shape file is collected from https://data.humdata.org/dataset/administrative-boundaries-of-bangladesh-as-of-2015

may involve talking with people who are dealing with poor health, with the social stigma attached to certain conditions and who are otherwise in potentially vulnerable situations' (p. 262). He also gave importance (1999) to recognizing the direct involvement in people's lives that qualitative research implies (see also, Cohen, Goto, Schreiber, & Torp-Pedersen, 2015; Kaley, Hatton, & Milligan, 2019). Qualitative studies are valuable because they provide insights that show us how conditions in particular places are thought to influence health and health-related behaviour, and they are powerfully suggestive of causal pathways relating environmental factors to individual health (Cummins, Curtis, Diez-Roux, & Macintyre, 2007; Green & Thorogood, 2004; Mykhalovskiy et al., 2008; Smyth, 2008). In developing the 'new geographies of HIV/AIDS' which are alternatives to the quantitative approach, many authors (i.e. Collins et al., 2016; Downing Jr., 2008; Ghosh, Wadhwa, & Kalipeni, 2009; Kuhanen, 2010; Lewis, 2016; Schatz, Madhavan, & Williams, 2011; Seckinelgin, 2012; Tobin, Cutchin, Latkin, & Takahashi, 2013) highlighted different issues such as women's vulnerability, the complex management of HIV/AIDS treatment, economic disparities, sexual culture change, the implications of the HIV research, bio-medical and behavioural models, place and social networks, the urban ecological approach and territorial stigma through insightful narratives.

Research in the geography of health needs to be sensitive to spatiality and temporality (Jones & Moon, 1993). Health research needs to be comprehensive enough to understand a complex interplay of forces, yet sufficiently focused to obtain valid and meaningful results. Following Gregory (1985), I have sought a methodology that recognizes that 'people make a difference and places make a difference' (p. 74). As qualitative methodologies reflect particular understandings of social life and meanings, the present research aims to develop an explanation of the social, health, geographical and policy aspects of HIV risk in Bangladesh. Here, qualitative data are used to understand the interactions between health risk and stigma, place and policy. In addition, the qualitative methods in this book examine how people lead their 'everyday lives' when they are marginalized or stigmatized in the society in the context of HIV vulnerability. The qualitative techniques for data collection include in-depth interviews, focus group discussions and participant observation, along with naturalistic observation of the risk sites. The grounded theory approach is used as a technique for analysing data.

In assessing the health, social and geographical influences on HIV/AIDS risk among vulnerable groups in Bangladesh, it is important to ask what data are needed and which methods should be adopted for data collection, analysis and theory building. As the present research objectives are to explore in-depth information regarding the challenges facing marginalized people in Bangladesh, this information is bound to be sensitive, covering many emotional and personal themes. Many of the stigmas, uncertainties, lack of awareness and unhygienic practices that place marginalized and vulnerable people's health at risk are difficult to quantify. However, much recent epidemiological research indicates that qualitative research is needed for insights into the risk behaviours of the most vulnerable population for HIV transmission and the circumstances of high-risk behaviours. Considering these factors, my study objectives have been best addressed through qualitative methods. Different qualitative techniques have been employed for this research for data collection and analysis. The survey design was organized as a data collection procedure to address the 'lifeworlds' of marginalized communities and also looking at the risk factors for HIV/AIDS infection. Data have been collected from both primary and secondary sources. At present, it looks as if very little health risk data regarding HIV/AIDS in Bangladesh exists, not only for the study areas but also for the whole country. The author therefore needed to collect his own data. Respondents were selected on the basis of a convenience sample design. All discussions used a semi-structured interview guide in the local dialect (in Bengali). Participants were given the choice of both time and venue for interviews or discussions. Photo Series 1.1 shows data collection with different people like CSW, DU, TW, local civil society people, peer educators in person and group.

During the field work, I worked with three social groups. First, I worked with people vulnerable to HIV infection: commercial sex workers (including hotel, residence, floating, brothel and street based), drug users (including heroin smokers and intravenous drug users), transport workers (including Indian and Bangladeshi truckers and rickshaw pullers) and female slum dwellers and people living with HIV (PLWH). Second, I talked with local elites (people in civil society): journalists,

NGO personnel, local government officials, international agency people, teachers, physicians, local elected representatives and religious persons. Both of these groups were interviewed in Jessore and Khulna. Finally, I worked with key personnel and policy planners for HIV in Dhaka, including Bangladeshi NGO high officials, international agency officials and members of funding organizations. I also joined in a round table discussion (RTD) with transport workers at Benapole. None of this was easy to do because no database exists at present on HIV/AIDS in the study areas and it was therefore difficult to anticipate the sampling frame and size of sample. It was necessary to select a sampling strategy for collecting the qualitative information. In terms of finding interviewees to take part in the study, participants were contacted through institutions or through snowballing. As the issue of HIV is a sensitive matter, I was very careful about the selection of my interviewees and FGD participants. I found some local NGOs who are working with vulnerable groups in Jessore and Khulna by using 'snowballing'. I took their help to get access to their stakeholders and also to discover and communicate with the respondents initially through their (NGO) peer educators, who worked at the field level, so that they could help me to find drug users, sex workers and transport workers in the 'open field'. Though I managed many interviews in the closed rooms of NGOs, others were arranged in the 'open field', like steamer terminals, tea stalls, transport terminals, residential streets, beside the highway or in rickshaw garages. I visited these different places on my own and used the snowballing technique to find them, particularly street sex workers and drug users. For instance, I discovered an addict in an area of Jessore that is known as a drug selling point. I introduced myself and we started our conversation in an open field sitting beside a coconut tree. Using the snowball sampling technique, I managed to develop relationships with some members of the marginalized community and find more participants from these groups. For transport workers, particularly Indian truckers, snowballing was less effective because nobody trusted me or wanted to speak about their personal lives in front of another friend. Here, a few former NGO peers assisted me to identify participants for interviewing and organizing an FGD. Firstly, these peers approached them to know whether they had spare time to spend talking to me. If anybody agreed, then I explained my purpose and importantly confirmed to him that I did not come from any intelligence branch, as most of them feared. After introductions, I boarded the driver's truck or sat beside the truck in the terminal area for the interview. Most days I used the afternoon period when they have free time. Most of them felt free to speak me when they observed that I did not ask their name. So, they trusted me and talked about their sexual life or drug habit which is usually considered to be a very secret matter. Sometimes they thought I was a doctor though I clearly stated my qualifications. For the FGD with Indian truckers, I was helped by an interpreter for the Hindi speaking drivers in Benapole terminal.

Apart from snowballing, I used NGO offices for interviews and for organizing FGDs with DUs, CSWs (mainly brothel-based) and PLWH. For example, in the brothel, I used some NGOs' offices inside the brothel as my interviewing place. When I requested NGO managers in the field level to suggest participants for my research, they chose them from different perspectives according to my instructions.

Talking with Indian Transport Workers

Talking with DU at Railstation

Talking with CSW at brothel setting

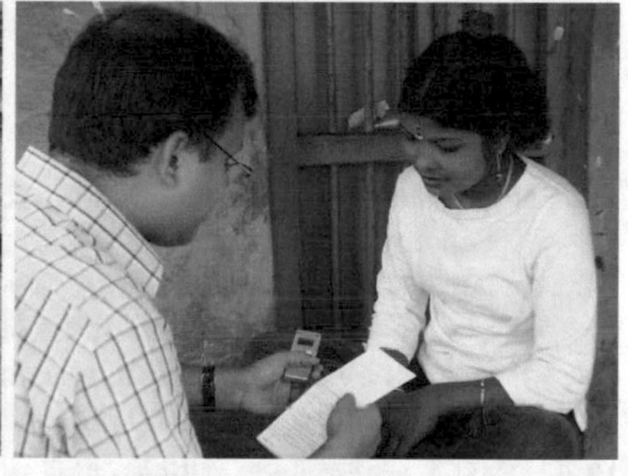

Discussion with DUs at a NGO office

Talking with DU at outside

Photo Series 1.1 Data collection with different people. (**a**) Talking with Indian Transport Workers. (**b**) Talking with CSW at her room. (**c**) Talking with DU at railstation. (**d**) Talking with CSW at brothel setting. (**e**) Discussion with DUs at a NGO office. (**f**) Talking with DU at outside. (**g**) Talking with CSW with her son and a NGO peer educator. (**h**) Talking with a local civil society member at Khulna

Most participants were not taking any visible benefits from that NGO, but had contact with the local NGO peers. Participants' ages ranged from 15 to 60 years. I wanted to carry out most interviews in the 'open field', but when I tried this I faced many problems from different corners. Most importantly, I observed that CSWs and

Talking with CSW with her son and a NGO peer educator

Talking with a local civil society member at Khulna

Fig. .2 (continued)

DUs are difficult to organize in a group format in the 'open field'. But they were in a controlled environment when I talked with them in a separate room of an NGO office. They were very attentive to my questions and articulate in their replies, which is absent in the 'open field'. And most importantly, due to the location, they were away from any physical threat. Both the inside and outside facilities provided by the NGOs enhanced the speed of my project. But there is a question that may arise out of an NGO's involvement with these participants for interviewing. Although I took the NGO's help, they had no influence on my research participants. I always assured my interviewees that I was not an NGO official, so they could easily share their views and I found this to be beneficial. In addition, many other prominent local people in both study areas and policy planners in Dhaka gave interviews in their offices. The grounded theory approach was extensively used in this research for analysing qualitative data. This approach was used as a form of field study that systematically applied procedural steps to develop an explanation about the social, geographical and policy aspects of HIV risk in Bangladesh. The vulnerable people's perceptions concerning the risk of HIV on their social and health conditions were fitted into a grounded theory approach in focusing on the realities of a situation. In this research, the goal of grounded theory is to seek a new concept that is compatible with the field evidence concerning HIV/AIDS risk. As the present study is a qualitative study and the data collected are exploratory and thematic in nature, there are no quantitative comparisons. This might have contributed to missing some more positive ideas in the research because the qualitative approach has some limits about the generality of the findings.

1.7 Concluding Remarks

HIV is one of the biggest social, economic and health challenges globally. After sub-Saharan Africa, south and south-east Asia is the most vulnerable to HIV/AIDS. Bangladesh, a South Asian country, is engaged in fighting and preventing major diseases such as diarrhoea and tuberculosis and in the contemporary period HIV/AIDS is regarded as one of the diseases most likely to put the country's population at risk. Bangladesh has many epidemiological and social factors that could produce a devastating epidemic and may lead the country into a health disaster. The country has experienced a low-level HIV epidemic for many years but the epidemic has recently entered the concentrated phase in the capital city. From the social and cultural point of view, as a Muslim nation, people do not widely perceive HIV/AIDS as a problem. Many believe that there is no scope for using opiates or promiscuous sex as Bangladeshis are chaste, pious and honest. In addition, they do not consider HIV/AIDS to be a threat to their conservative society, though poverty, discrimination, illiteracy, cross border trade, high population movements and unequal health access are very common. The transmission of HIV/AIDS is related to sexual and drug use behaviour, which is a domain of privacy and secrecy. Initial reaction to open discussion of sexual and opiate drug issues is characterized by fear, stigma and discrimination, leading to social rejection. It is commonly believed that HIV/AIDS is still underreported in Bangladesh. People do not usually want to disclose their status and take treatment for fear of shame and embarrassment, or loss of social and economic security. Along with PLWH, sex workers and drug users are considered to be the most at-risk group for HIV, and they have to bear a very bad image in the community and are often ostracized. Above all, these social repercussions and cultural barriers in risk prevention affect all aspects of the HIV/AIDS situation in conservative Bangladesh, which is a contribution to increasing the level of risk and vulnerability. The global AIDS pandemic is recognized worldwide as a substantial public health problem. This book has been planned to look at the current threat of HIV and its adverse health and social effects, as well as public policy and practice in Bangladesh.

References

Akhter, S., Anwar, I., Akter, R., Kumkum, F. A., Nisha, M. K., Ashraf, F., et al. (2016). Barriers to timely and safe blood transfusion for PPH patients: Evidence from a qualitative study in Dhaka, Bangladesh. *PLoS One, 11*(12), e0167399.

Alam, M. Z., Hossain, M. A., & Chowdhury, M. A. (2015). Prevalence and risk factors of Human Immunodeficiency virus, Hepatitis B virus, Hepatitis C virus and Syphilis infections among nonprofessional blood donors in Chittagong, Bangladesh. *Asian Journal of Medical and Biological Research, 1*(3), 518–525.

Alam, N., Chowdhury, M. E., Mridha, M. K., Ahmed, A., Reichenbach, L. J., Streatfield, P. K., et al. (2013). Factors associated with condom use negotiation by female sex workers in Bangladesh. *International Journal of STD & AIDS, 24*(10), 813–821.

Andrews, G. J., & Evans, J. (2008). Understanding the reproduction of health care: Towards geographies in health care work. *Progress in Human Geography, 32*(6), 759–780.
Azim, T., Islam, M. N., Bogaerts, J., Mian, M. K., Sarker, M. S., Fattah, K. R., et al. (2000). Prevalence of HIV and Syphilis among high-risk groups in Bangladesh. *AIDS, 14*(2), 210–211.
Azim, T., Rahman, M., Alam, M. S., Chowdhury, I. A., Khan, R., Reza, M., et al. (2008). Bangladesh moves from being a low-prevalence nation for HIV to one with a concentrated epidemic in injecting drug users. *International Journal of STD & AIDS, 19*(5), 327–331.
Aziz, K. M. A. and Maloney, C. (1985). Life stages, gender and fertility in Bangladesh. Monograph no. 3, ICDDR'B, Dhaka, Bangladesh
Bean, P. (2002). *Drugs and crime*. Portland, OR: Willan Publishing.
Bhuiya, A., Hanifi, S. M. A., Hossain, M., & Aziz, A. (2000). Effects of an AIDS awareness campaign on knowledge about AIDS in a remote rural area of Bangladesh. *International Quarterly of Communication Health Education, 19*, 51–63.
Bosu, A. (2013). Sexual act with multiple sex partners is an upcoming threat for rapid transmission of HIV among People Who Inject Drugs (PWID) in Bangladesh. *Sexually Transmitted Infections, 89*(Suppl 1), A179–A179.
Caldwell, B., & Pieris, I. (1999). Continued high-risk behaviour among Bangladeshi males. In J. C. Caldwell, P. Caldwell, J. Anarfi, et al. (Eds.), *Resistance to behavioural change to reduce HIV/AIDS infection in predominantly heterosexual epidemics in third world countries* (pp. 183–196). Canberra, Australia: Health Transition Center, Australian National University.
Campbell, C., & Williams, B. (1999). Beyond the biomedical and behavioural: Towards an integrated approach to HIV prevention in the Southern African mining industry. *Social Science and Medicine, 48*(11), 1625–1639.
Chan, P. A., & Khan, O. A. (2007). Risk factors for HIV infection in males who have sex with males (MSM) in Bangladesh. *BMC Public Health, 7*, 153.
Chelala, C. and Beyrer, C. (1999, September) Drug use and HIV/AIDS in Burma. *The Lancet, 354*
Choudhury, A. Y., Arjumand, L., & Piwoo, J. S. (1997). *A rapid assessment of health seeking behaviour in relation to sexually transmitted disease*. Dhaka, Bangladesh: PIACT Bangladesh.
Cohen, A. T., Goto, S., Schreiber, K., & Torp-Pedersen, C. (2015). Why do we need observational studies of everyday patients in the real-life setting? *European Heart Journal Supplements, 17*(Suppl. D), D2–D8.
Collins, A. B., Parashar, S., Closson, K., Turje, R. B., Strike, C., & McNeil, R. (2016). Navigating identity, territorial stigma, and HIV care services in Vancouver, Canada: A qualitative study. *Health and Place, 40*(July), 169–177.
Cummins, S., Curtis, S., Diez-Roux, A. V., & Macintyre, S. (2007). Understanding and representing 'place' in health research: A relational approach. *Social Science and Medicine, 65*(9), 1825–1838.
Downing Jr., M. J. (2008). The role of home in HIV/AIDS: A visual approach to understanding human-environment interactions in the context of long-term illness. *Health & Place, 14*(2), 313–322.
Dunn, C. E. (2007). Participatory GIS: A people's GIS? *Progress in Human Geography, 31*(5), 616–637.
Folmar, S., Alam, S. M. N., & Raihan Sharif, A. H. M. (1992). *Condom use in Bangladesh*. Final Report, Family Health International, Dhaka.
Ford, N., Siregar, K., Ngatimin, R., & Maidin, A. (1997). The hidden dimension: Sexuality and responding to the threat of HIV/AIDS in South Sulawesi, Indonesia. *Health & Place, 3*(4), 249–258.
Gazi, R., Khan, S. I., Haseen, F., Sarma, H., Islam, M. A., Wirtz, A. L., et al. (2009). Young clients of hotel-based sex workers in Bangladesh: Vulnerability to HIV, risk perceptions, and expressed needs for interventions. *International Journal of Sexual Health, 21*(3), 167–182.
Gazi, R., Mercer, A., et al. (2008). An assessment of vulnerability to HIV infection of boatmen in Teknaf, Bangladesh. *Conflict and Health, 2*(5), 1–11.

Ghosh, J., Wadhwa, V., & Kalipeni, E. (2009). Vulnerability to HIV/AIDS among women of reproductive age in the slums of Delhi and Hyderabad, India. *Social Science and Medicine, 68*(4), 638–642.

Gibney, L., Choudhury, P., Khawaja, Z., Sarker, M., Islam, N., & Vermund, S. H. (1999). HIV/AIDS in Bangladesh: An assessment of biomedical risk factors for transmission. *International Journal of STD & AIDS, 10*, 338–346.

Gibney, L., Choudhury, P., Khawaja, Z., Sarker, M., & Vermund, S. H. (1999). Behavioural risk factors for HIV/AIDS in a low-HIV prevalence Muslim nation, Bangladesh. *International Journal of STD & AIDS, 10*, 186–194.

GOB. (2002). *National assessment of situation and responses to opiate use in Bangladesh, Department of Narcotics control, Joint survey conducted by government and non-government agencies*. Dhaka, Bangladesh: Government of Bangladesh (GOB).

GOB. (2004). *The counter trafficking framework report: Bangladesh perspective*. Dhaka, Bangladesh: Ministry of Women and Children Affairs, Government of Bangladesh (GOB).

Green, J., & Thorogood, N. (2004). *Qualitative methods for health research*. London, UK: Sage.

Gregory, D. (1985). People, places and practices: The future of human geography. In R. King (Ed.), *Geographical futures*. Sheffield, UK: Geographical Association.

Haque, M. A., Hossain, M., Chowdhury, M., & Uddin, M. J. (2018). Factors associated with knowledge and awareness of HIV/AIDS among married women in Bangladesh: Evidence from a nationally representative survey. *Journal of Social Aspects of HIV/AIDS Research Alliance, 15*(1), 121–127.

Hasan, M. (2007). *Harassment pattern of sex workers in Bangladesh: A situational analysis of three brothels, narratives and perspective in sociology understanding the past, envisaging the future*. Proceedings of the 8th annual conference of Hong Kong sociological association, Hong Kong Shue Yan University.

Haseen, F., Chawdhury, F. A. H., Hossain, M. E., Huq, M., Bhuiyan, M. U., Imam, H., et al. (2012). Sexually transmitted infections and sexual behaviour among youth clients of hotel-based female sex workers in Dhaka, Bangladesh. *International journal of STD & AIDS, 23*(8), 553–559.

Hosain, G. M. M., & Chatterjee, N. (2005). Beliefs, sexual behaviours and preventive practices with respect to HIV/AIDS among commercial sex workers in Daulatdia. *Bangladesh, Public Health, 119*, 371–381.

Hossain, K. J., Akter, N., Kamal, M., Mandal, M. C., & Aktharuzzaman, M. (2012). Screening for HIV among substance users undergoing detoxification. *International journal of STD & AIDS, 23*(10), e1–e5.

Hossain, M., Mani, K. K., Sidik, S. M., Shahar, H. K., & Islam, R. (2014). Knowledge and awareness about STDs among women in Bangladesh. *BMC Public Health, 14*, 775.

Hossain, M. B., & Kippax, S. (2011). Stigmatized attitudes toward people living with HIV in Bangladesh: Health care workers' perspectives. *Asia Pacific Journal of Public Health, 23*(2), 171–182.

Hossain, S. M. I., Bhuiya, I. and Streatfield, K. (1996). *Professional blood donors, blood banks and risk of STDs and HIV/AIDS: A study in selected areas in Bangladesh* (Regional working papers, South and East-Asia, No. 6). Dhaka, Bangladesh: The Population Council.

IFRC. (2018). *World Disasters Report: Focus on HIV/AIDS*. Geneva, Switzerland: International Federation of Red Cross and Red Crescent Societies (IFRC).

IOM. (2004). *Revisiting the human trafficking paradigm: The Bangladesh experience, International Organization for Migration (IOM), The Bangladesh thematic group on trafficking*. Geneva, Switzerland: IOM.

Islam, M. T., Mostafa, G., Bhuiya, A. U., Hawkes, S., & de Francisco, A. (2002). Knowledge on, and attitude toward, HIV/AIDS among staff of an international organization in Bangladesh. *Journal of Health, Population and Nutrition, 20*(3), 271–278.

Islam, S. (1981). *Indigenous abortion practices in rural Bangladesh*. Dhaka, Bangladesh: Women for Women Research and Study Group.

Islam, S. K., Hossain, K. J., Kamal, M., & Ahsan, M. (2003). Prevalence of HIV infection in the drug addicts of Bangladesh: Drug habit, sexual practice and lifestyle. *International Journal of STD & AIDS, 14*(11), 762–764.

Islam, S. M. S., Biswas, T., Bhuiyan, F. A., Islam, M. S., Rahman, M. M., & Nessa, H. (2015). Injecting drug users and their health seeking behaviour: A cross-sectional study in Dhaka, Bangladesh. *Journal of Addiction, 2015*, 8 pages.

Jenkins, C. (1999). Resistance to condom use in a Bangladesh brothel. In J. C. Caldwell, P. Caldwell, J. Anarfi, et al. (Eds.), *Resistance to behavioural change to reduce HIV/AIDS infection in predominantly heterosexual epidemics in third world countries* (pp. 211–222). Canberra, Australia: Health Transition Center, Australian National University.

Jones, K., & Moon, G. (1993). Medical geography: Taking space seriously. *Progress in Human Geography, 17*(4), 515–524.

Kaley, A., Hatton, C., & Milligan, C. (2019). Health geography and the 'performative' turn: Making space for the audio-visual in ethnographic health research. *Health and Place, 60*(November), 102210.

Kamal, S. M., Hassan, C. H., & Salikon, R. H. (2015). Safer sex negotiation and its association with condom use among clients of female sex workers in Bangladesh. *Asia Pacific Journal of Public Health, 27*(2), NP2410–NP2422.

Karim, M., & Mona, N. (2013). Knowledge attitude practises about sexually transmitted disease among the commercial sex workers. *Sexually Transmitted Infections, 89*(Suppl. 1), A204–A204.

Kearns, R. A. (1995). Medical geography: Making space for difference. *Progress in Human geography, 19*(2), 251–259.

Khan, M. A., Rahman, M., Khanam, P. A., Barkat-e-Khuda, K. T. T., & Ashraf, A. (1997). Awareness of sexually transmitted disease among women and service providers in rural Bangladesh. *International Journal of STD & AIDS, 8*, 688–696.

Khan, M. H. (1999). *Collection of behavioral and socio-economic data from selected population of university students through rapid Assessment for use in program on HIV/AIDS/STDs*. Dhaka, Bangladesh: AIDS Awareness Foundation (AAF).

Khan, M. M. K. (2006). *Demand reduction approach and social impediments to solution of drug problem: A review in the context of Bangladesh, In the publication of Department of Narcotics Control, on the International day against drug abuse and illicit trafficking*. Dhaka, Bangladesh: Government of Bangladesh.

Khan, S. I., Hudson-Rodd, N., Saggers, S., & Bhuiya, A. (2005). Men who have sex with men's sexual relations with women in Bangladesh. *Culture Health and Sexuality, 7*, 159–169.

Khanam, R., Reza, M., Ahmed, D., Rahman, M., Alam, M. S., Sultana, S., et al. (2017). Sexually transmitted infections and associated risk factors among street-based and residence-based female sex workers in Dhaka, Bangladesh. *Sexually Transmitted Diseases, 44*(1), 22–29.

Knight, V. C. (2006). *Transport workers at risk to HIV: Documenting our experience 2000–2004, HIV program*. Dhaka, Bangladesh: Care Bangladesh.

Kuhanen, J. (2010). Sexualised space, sexual networking & the emergence of AIDS in Rakai, Uganda. *Health and Place, 16*(2), 226–235.

Lewis, N. M. (2016). Urban encounters and sexual health among gay and bisexual immigrant men: Perspectives from the settlement and aids service sectors. *Geographical Review, 106*(2), 235–256.

MacDonald, D., & Mansfield, D. (2001). Drugs and Afghanistan. *Drugs education: Prevention and Policy, 8*(1), 1–16.

Mac-Phail, C., & Campbell, C. (2001). 'I think condoms are good, but, aai, I hate those things': Condom use among adolescents and young people in a South African township. *Social Science and Medicine, 52*(11), 1613–1627.

Mahmood, S. A. I. (2007). Leadership is a must to combat this scourge, *The Daily Star*, December 1, Dhaka.

Maloney, C., Aziz, K., & Sarker, P. (1981). *Beliefs and fertility in Bangladesh*. ICDDR'B, Dhaka.

Mann, J. M. (1987). The global AIDS situation. *World Health Statistics Quarterly, 40*, 185–192.

Mason, M. (2007). *Applications of geographic information systems to community psychology research & practice*. Discussant paper presented at the annual meeting of the society for community research and action, Pasadena, CA.

Mercer, A., Khanam, R., Gurley, E., & Azim, T. (2007). Sexual risk behavior of married men and women in Bangladesh associated with husbands' work migration and living apart. *Sexually Transmitted Diseases, 34*(5), 265–273.

Mollah, A. H., Siddiqui, M. A., Anwar, K. S., Rabbi, F. J., Tahera, Y., Hassan, M. S., et al. (2004). Seroprevalence of common transfusion-transmitted infections among blood donors in Bangladesh. *Public Health, 118*(4), 299–302.

Mondal, N. I., Takaku, H., Ohkusa, Y., Sugawara, T., & Okabe, N. (2009). HIV/AIDS acquisition and transmission in Bangladesh: Turning to the concentrated epidemic. *Japanese Journal of Infectious Disease, 62*(2), 111–119.

Mowla, M. R., & Sattar, M. A. (2016). Recent trends in sexually transmitted infections: The Chittagong, Bangladesh experience. *Sexually Transmitted Infections, 92*(5), 349–349.

Muntasir, Z. A. (2005). *Reduce injecting drug use and prevent the spread of HIV in Dhaka city, In the publication of Department of Narcotics Control, on the International day against drug abuse and illicit trafficking*. Dhaka, Bangladesh: Government of Bangladesh.

Mykhalovskiy, E., Armstrong, P., Armstrong, H., Bourgeault, I., Choiniere, J., Lexchin, J., et al. (2008). Qualitative research and the politics of knowledge in an age of evidence: Developing a research-based practice of immanent critique. *Social Science and Medicine, 67*(1), 195–203.

NASP. (2004). *Bangladesh country profile on HIV and AIDS, National AIDS/STD Programme (NASP), Ministry of Health and Family welfare*. Dhaka, Bangladesh: Government of Bangladesh.

NASP. (2005). *National strategic plan for HIV/AIDS 2004–2010, National AIDS/STD Programme (NASP), Ministry of Health and Family welfare*. Dhaka, Bangladesh: Government of Bangladesh.

NASP. (2009). *Population size estimates for most at risk populations for HIV in Bangladesh, National AIDS/STD Program, Directorate General of Health Services, Ministry of Health and Family Welfare*. Dhaka, Bangladesh: Government of Bangladesh.

NASP. (2014). *Revised 3rd national strategic plan for HIV and AIDS response 2011–2017, National AIDS/STD Program, Ministry of Health and Family Welfare*. Dhaka, Bangladesh: Government of Bangladesh.

NASP. (2016). *Fourth National Strategic Plan for HIV and AIDS response, National AIDS/STD Program, Ministry of Health and Family Welfare*. Dhaka, Bangladesh: Government of Bangladesh.

Nessa, K., Waris, S. A., Alam, A., Huq, M., Nahar, S., et al. (2005). Sexually transmitted infections among brothel-based sex workers in Bangladesh: High prevalence of asymptomatic infection. *Sexually Transmitted Diseases, 32*(1), 13–19.

Panos. (2006). *Keeping the promise? A study of progress made in implementing the UNGASS declaration of commitment on HIV/AIDS in Bangladesh*. Dhaka, Bangladesh: The Panos Global AIDS Programme.

Paul, A. (2009) Geographies of HIV/AIDS in Bangladesh: Vulnerability, Stigma and Place. Durham theses, Durham University. http://etheses.dur.ac.uk/1348/

Paul, A. (2019). Geographies of HIV/AIDS in Bangladesh: Global perspectives local contexts and research gaps. In S. T. Islam & A. Paul (Eds.), *Geography in Bangladesh: Concepts methods and applications* (pp. 54–73). London, UK: Routledge, Taylor and Francis.

Paul, A., Atkins, P. J., & Dunn, C. E. (2012). Borders and HIV risk: A qualitative investigation in Bangladesh. *Oriental Geographer, 53*(1), 73–82.

Rahman, M., & Zaman, M. S. (2005). Awareness of HIV/AIDS and risky sexual behaviour among male drug users of higher socioeconomic status in Dhaka, Bangladesh. *Journal of Health, Population and Nutrition, 23*(3), 298–301.

Rahman, M. S., & Rahman, M. L. (2007). Media and education play a tremendous role in mounting AIDS awareness among married couples in Bangladesh. *AIDS Research and Therapy, 4*(1), 1.

Rahman, M., Alam, A., Nessa, K., Hossain, A., et al. (2000). Etiology of sexually transmitted infections among street-based female sex workers in Dhaka, Bangladesh. *Journal of Clinical Microbiology, 38*(3), 1244–1246.

Human Rights Watch Report. (2003). *Ravaging the vulnerable: Abuse against persons at high risk of HIV infection*, August 19, USA

Rob, U., & Mutahara, M. U. (2000). Premarital sex among urban adolescents in Bangladesh. *International Quarterly of Community Health Education, 20*(1), 103–111.

Roy, T., Anderson, C., Evans, C., & Rahman, M. S. (2010). Sexual risk behaviour of rural-to-urban migrant taxi drivers in Dhaka, Bangladesh: A cross-sectional behavioural survey. *Public Health, 124*(11), 648–658.

Sabin, K., Hawkes, S., Rahman, M. et al (1997) *Barriers to seeking treatment for sexually transmitted diseases among the Dhaka slum dwellers.* ASCON VI Conference, International Centre for Diarrhoeal Disease Research, Bangladesh, (ICDDR, B), March, Special Publication No. 57, Dhaka.

Sarafian, I. (2012). Process assessment of a peer education programme for HIV prevention among sex workers in Dhaka, Bangladesh: A social support framework. *Social Science & Medicine, 75*(4), 668–675.

Sarkar, K., Bal, B., Mukherjee, R., Chakraborty, S., Saha, S., Ghosh, A., et al. (2008). Sex-trafficking, violence, negotiating skill, and HIV infection in brothel-based sex workers of eastern India, adjoining Nepal, Bhutan, and Bangladesh. *Journal of Health Population and Nutrition, 26*, 223–231.

Sarkar, S., Islam, N., Durandin, F., et al. (1998). Low HIV and high STD among commercial sex workers in a brothel in Bangladesh: Scope for prevention of larger epidemic. *International Journal of STD & AIDS, 9*, 45–47.

Sarma, H., & Oliveras, E. (2011). Improving STI counselling services of non-formal providers in Bangladesh: Testing an alternative strategy. *Sexually transmitted infections, 87*(6), 476–478.

Scambler, G., & Paoli, F. (2008). Health work, female sex workers and HIV/AIDS: Global and local dimensions of stigma and deviance as barriers to effective interventions. *Social Science & Medicine, 66*(8), 1848–1862.

Schatz, E., Madhavan, S., & Williams, J. (2011). Female-headed households contending with AIDS-related hardship in rural South Africa. *Health & place, 17*(2), 598–605.

Seckinelgin, H. (2012). The global governance of success in HIV/AIDS policy: Emergency action, everyday lives and Sen's capabilities. *Health and Place, 18*(3), 453–460.

Smyth, F. (2008). Medical geography: Understanding health inequalities. *Progress in Human Geography, 32*(1), 119–127.

Sui, D. Z. (2007). Geographic information systems and medical geography: Toward a new synergy. *Geography Compass, 1*(3), 556–582.

The Independent. (2017). Illegal blood banks, an editorial. *The Independent,* 5th March, Dhaka.

Tobin, K. E., Cutchin, M., Latkin, C. A., & Takahashi, L. M. (2013). Social geographies of African American men who have sex with men (MSM): A qualitative exploration of the social, spatial and temporal context of HIV risk in Baltimore, Maryland. *Health & place, 22*(July), 1–6.

Townley, G., Kloos, B., & Wright, P. A. (2009). Understanding the experience of place: Expanding methods to conceptualize and measure community integration of persons with serious mental illness. *Health & Place, 15*(2), 520–531.

UNAIDS. (2006). *UNAIDS Annual report: Making the money work*. Geneva, Switzerland: Joint United Nations Programme on HIV/AIDS (UNAIDS).

UNAIDS. (2012). *Report on the global AIDS epidemic*. Geneva, Switzerland: Joint United Nations Programme on HIV/AIDS (UNAIDS).

UNAIDS. (2017). *UNAIDS data 2017*. Geneva, Switzerland: Joint United Nations Programme on HIV/AIDS (UNAIDS).

UNAIDS. (2018). *Miles to go: Closing gaps breaking barriers righting injustices*. Geneva, Switzerland: Joint United Nations Programme on HIV/AIDS (UNAIDS).

WHO. (1997). *AIDS: Some questions and answers, World Health Organization (WHO)*. New Delhi, India: Regional Office for South-East Asia.

WHO. (2019). Fact Sheets: HIV/AIDS, 15 November, World Health Organization (WHO), https://www.who.int/news-room/fact-sheets/detail/hiv-aids

Wilton, R. D. (1999). Qualitative health research: Negotiating life with HIV/AIDS. *The Professional Geographer, 51*(2), 254–264.

World Bank Report. (2006). *AIDS in South Asia: Understanding and responding to a heterogeneous epidemic*. Launched at the 16th international AIDS conference in Toronto, Canada.

Yaya, S., Bishwajit, G., Danhoundo, G., Shah, V., & Ekholuenetale, M. (2016). Trends and determinants of HIV/AIDS knowledge among women in Bangladesh. *BMC Public Health, 16*(1), 812.

Yiannakoulias, N., Svenson, L. W., & Schopflocher, D. P. (2005). Diagnostic uncertainty and medical geography: What are we mapping? Commentary. *The Canadian Geographer, 49*(3), 291–300.

Chapter 2
'Lifeworlds' of Marginalized People

2.1 Introduction

The term 'lifeworld' ('lebenswelt') was used by Habermas (1987) to describe the collection of behaviours, expectations, norms and communicative acts (Wilkie, 2001) that comprise everyday life and serve to link individuals (Barclay, 2020; Mau, 2010; Stone, 2013). The study of everyday life in social research involves 'the necessity of subjecting one's own activities to practical knowledge and routines whose heterogeneity and lack of systemicity is rarely theorized' (Featherstone, 1995; p. 55). Theorists such as Schutz and Berger and Luckman recast the individual as an active agent in the construction of meaning in everyday life (Bennett, 2005) in their social context or meanings (Morgan, 2004; Travis, 2017). In this approach, the significance of everyday life is inseparable from the indexical meanings ascribed to it by individual actors (Gardiner, 2000; Marques, 2019). This interpretation of everyday life is further developed by Erving Goffman through his application of a dramaturgical model of everyday interactions (Bennett, 2005). Thus, Goffman (1959) argues through gaining a 'practical experience' of everyday life characterized by the internalization of social roles, individuals also learn how to manage and negotiate those roles through the creation of 'front-stage' and 'backstage' selves. In doing so, individuals' creativity manipulates the everyday, making it tolerable through creating spaces for the subversion of conformity. Goffman's approach to the everyday suggests an inventory of performances spatially arranged across the geography of everyday life (Highmore, 2002; Karner, 2007). Chaney (2002) assumes everyday life to be the forms of life we routinely consider unremarkable and thus take for granted. The reality of habitual experience is provided in the routines or rhythms of occupations, relationships and residences.

Geographers are of interest as they focus on the individual, the short term and small-scale, pointing out the routine nature of everyday life, a 'chronogeographical' analysis pioneered by the Lund geographers (Holloway & Hubbard, 2001). In a

A. Paul, *HIV/AIDS in Bangladesh*, Global Perspectives on Health Geography,
https://doi.org/10.1007/978-3-030-57650-9_2

writing, John Eyles (1989), on his 'the geography of everyday life' follows similar arguments in suggesting that it is desirable that geographer examine the everyday. Though he contends that such a sensitivity to the common place is necessary for the development of a human-centred geography, few (for example, May, 1996; Watts, 1991) could dispute that geographical analysis of the everyday offers a way of thinking about the relationships between individuals and their surroundings in a period of rapid time-space compression. Seamon (2015) was concerned with how individuals' phenomenological immersion into a geography of everyday life constructed their 'lifeworlds'. To explore people's 'lifeworlds', he argued that it is necessary to consider the relationships between people's behaviour (what people do in places) and their experiences of place (Holloway & Hubbard, 2001).

In the 1990s, geographers began to make contributions to an understanding of the AIDS epidemic in the form of mapping to illustrate the origin of the HIV and its diffusion at global, national and local levels. These contributions of spatial diffusion of the virus in the geography of AIDS are valuable and important but provide only a 'single and partial geography of AIDS' (Wilton, 1996; p. 70). Relatively, little attention has been paid to the socio-cultural, economic and political dimensions of the disease or to the ways in which individual risk behaviours are related to non-spatial processes, in particular geographical settings (Asthana, 1998; Craddock, 2000). One promising response to this research gap is Wilton's (1996) ethnographic study of the daily life experiences of Los Angeles men with symptomatic HIV/AIDS that disaggregates the social, physical and psychological dimensions that help him to show how place matters in the lives of these men. Here, space is important in terms of material conditions for people living with HIV/AIDS in the development of their daily paths and their shifting social networks, where people's daily worlds and routines changed after diagnosis. In his subsequent paper, Wilton gives a framework of qualitative research with a critique (Wilton, 1999). Brown (1995) also carried out ethnographic work and argued that geographers' preoccupation with the spatial diffusion of the virus threatens to reduce gay men's bodies to biological hosts, a subject of interest only to the extent that they are able to transport HIV across space. In another piece of empirical work that included a distinctively ethnographic approach to health, Brown (1997) looked at the local politics of AIDS and tracked the location of political engagement with the AIDS crisis. Craddock (2000) proposed a framework which combines an approach to mapping vulnerability with feminist and post-structural approaches that focus more attention upon the role of social identities and cultural framings of diseases particularly calculations of disease risk. A combination of these approaches would result in a more effective framework for evaluating vulnerability and subsequently for generating effective disease prevention strategies. Browne and Barrett (2001) discussed the 'moral geographies' in different campaigns in the context of HIV prevention in Africa. They have raised the question of how states define and represent norms of sexual morality and, in this case, the AIDS pandemic has forced states to redraw the 'moral boundaries' of consideration and acceptability for public consumption.

The evolution of the HIV/AIDS pandemic has shown a consistent pattern through which marginalization of individuals and groups heighten their vulnerability to HIV exposure. Marginalization engages the formative influence of economics and politics in contemporary life, including their increasing interaction with the environment (Atherton, 2003; Lorway et al., 2018). This chapter looks at the everyday geography of marginalized people's daily rituals, physical discomfort, sorrows and anger, dreams and frustrations, all of which fuel their vulnerability. In order to know how marginalized people lead their 'everyday lives', here I consider brothel and non-brothel sex workers, opiate users and people living with HIV as marginalized groups because of their negative status in and neglect by mainstream society, and discrimination by the state. I have tried to hear the distressful 'stories' of their life. From these stories, I have reconstructed their everyday rituals, physical vulnerability and frustrations.

2.2 Female Sex Workers' Everyday Life and Rituals

The centrality of women's role in everyday life, ironically, gets intertwined into the fabric of social relations and social structures in ways that create women's subordination and unequal position (Heyzer, 1986), especially in patriarchal societies like Bangladesh where life expectancy for women is lower than for men (Kumar, Raju, Atkins, & Townsend, 1997). Men are usually dominant in the allocation of scarce resources and this structured inequality has a major impact on women's health (Doyal, 1995; Khan & Khanum, 2000; McNeil, Shannon, Shaver, Kerr, & Small, 2014). Poverty, illiteracy and traditional customs often make women vulnerable to HIV infection (UNAIDS, 2002), and this has now become a major threat to women (Lorway et al., 2018), especially those involved in the commercial heterosex industry around the world (Guha, 2017) including brothel. Literally, a brothel is 'a house of prostitutes' where clients can view, select, arrange and execute a transaction with prostitutes (Davidson, 2006). The term 'sex worker' is sometimes used in place of the word 'prostitute' who is exploited sometimes for money in a commercial sex business (Farley, 2001) and depends on their own labour (Patterson, 2015). Prostitution legitimizes commercial sexual exploitation (Gutierrez-Garza, 2013) but offers no dignity or safety to the person in prostitution (Bungay, Halpin, Atchison, & Johnston, 2011; Farley, 2001). Women's entry into prostitution is characterized by an act of resistance to the experience of relative poverty or threat of it. Prostitution, a worldwide phenomenon (Pullen, 2005) is often a response to poverty, financial hardship, need, homelessness and unemployment (Kempadoo & Doezema, 1998; Lorraine van Blerk, 2016), and a result of women's marginalization (Fassi, 2011; Grenfell, Platt, & Stevenson, 2018). A study by Chant and Mcllwaine (1995) also shows that women often have to turn to sex work in order to support their children, and some are in the position of having to help parents and siblings as well (Rawat & Kumar, 2015). Since, prostitution is a direct outcome of women's gender subordination (Gutierrez-Garza, 2013), such links should be recognized by AIDS programmes, and ways found to combat such traditions (Patterson, 2015).

Though prostitution is prohibited in many countries of the world, some sex workers work independently or maintain link with the clients or manager in Bangladesh. Khan and Arefeen (1989) found that the majority of prostitutes were not abducted but had chosen the profession in the face of limited alternative options and perceived the trade to have major economic advantages. Khan (1988) and the UBINIG Report (1995) also show that a network of brokers and middlemen exists throughout Bangladesh who tempt girls and women from poor families with the promise of lucrative jobs, and it seems that dowry-related violence is a push factor (Barton, 1999). Hammond (2008) mentioned that poverty and other social and cultural factors related to gender-based exploitation are the main reasons for entering into prostitution, especially as bonded sex workers in the brothel (Tahmina & Moral, 2004; Taposh, 2006). There are a few 'registered' brothels where many girls and their family members reside. Due to legitimacy dilemma, they need an official document called a 'license'. The number of sex workers in Bangladesh is difficult to judge. There are estimates ranging from 150,000 to 200,000 brothels, streets, hotels and residences (see also, Gibney, Choudhury, Khawaja, Sarker, & Vermund, 1999). Brothel closure in many locations is one important reason, along with social and economic factors, for the recent explosion in the numbers of non-brothel-based sex workers (Ara, 2005; Bloem et al., 1999; Jenkins & Rahman, 2002). Sex work itself may have a negative impact on health (Seib, Fischer, & Najman, 2009). Thus, without legal protection (UNDP, 2012) under Bangladeshi law, girls and women involved in this profession have to struggle daily for their survival (Ahsan, Ahmad, Eusuf, & Roy, 1999; Ullah, 2005). Such uncertainty including their dreams and aspirations puts such girls in a more vulnerable situation with regard to STDs and HIV (Khanam et al., 2017). This section will look at how sex workers lead their lives at the brothel along with their different rituals. Some photos of the brothel-based sex workers are shown in Photo Series 2.1 as symbolic. Here, bonded sex workers who put their body at risk as they have the least freedom in the brothel setting will be described. It will also touch on those aspects that are playing a role in the uncertainties of brothel and non-brothel sex workers' income and physical discomfort.

Life at the Brothel Brothel-based sex workers, a marginalized community as they are socially isolated, lead everyday lives that are very different from those of 'straight' women in Bangladesh. A brothel girl's life is a matter of everyday 'business' with common routines starts early in the morning and continues until late evening, with some breaks. After taking a shower and breakfast they go to the 'gate' for a customer. Their rule is that they have to stand in line so that everyone will eventually get a customer, although some girls have regulars. In many brothels, girls take their main food (lunch and dinner) in a mess system and live in a house together sharing their rent, food costs, electricity bills and night guard bills. When a customer comes the others wait outside. On average, each gets 2–5 customers during the day, but customers are not allowed to stay at night in most brothels. Usually the girls do not go outside except in emergencies. In such an inward-looking environment, quarrels are common and sometimes turn into physical assault. Brothel sex workers' main form of entertainment is Bangladeshi and Indian romantic movies.

Photo Series 2.1 Sex worker's everyday life at brothel

Moreover, in almost every house they have a cassette or CD player to listen to music. One Fultola girl, Nazma told me about her everyday life:

> 'I get up in the morning at 7.30, then go to the toilet and wash my face, clean the house, and collect water and take a shower. Then I eat and afterwards go to the gate and collect a customer. After a lunch of fish and vegetables I take a rest. A customer may come in the meantime, and if he says my name at the gate then he will be able to come to me. If the customer comes after lunch, then I will not take a rest. In the evening I roam around in the gate area for more customers, chatting with other girls, then come to my room and cook something and watch TV. After taking dinner, I go to sleep. It's everyday life. I take a break when it is my 'period', and at that time if any customer comes for me then I will give him to another girl and collect some money'.

Many brothel sex workers want their own or adopted children with them, partly to satisfy their emotional needs for attachment and partly to ensure an income when aged. I found that many brothel sex workers keep their children with them, although it seems that there are psychological effects due to the mother's profession. In addition, the children's presence can cause embarrassment and sometimes hampers earnings. In Baniashanta brothel, Santa told me that she has been facing a problem with her little son, who becomes upset. She says '*sometimes my son shows anger when I talk to any man or any customer comes to my house and he stands at the closed door and cries, and asks me to open the door. Sometimes the other girls help me by taking him to their room*'. Some sex workers send their babies to an NGO shelter house, as they cannot give the appropriate care. Some girls like Parul keep her child in a house in the adjacent village and pays monthly for food and board.

When comparing their present life with the past, many girls become very emotional and nostalgic. In Baniashanta, one former sex worker, Nargis told me that she had enjoyed her life with many pleasures like drinking alcohol with customers, songs and dancing every weekend but now the flow of customers is less and there is no alcohol shop in the brothel. She thinks that previously the brothel was a place of enthusiasm and girls were affectionate to each other but now nobody cares about the others. During religious festivals like Ramadan or Eid, most of the brothel girls usually stay indoors. Although most do not follow religious fasting due to their own sense of 'impurity', they try to celebrate the festivals by cooking special dishes, wearing good clothes and sharing food with their 'private' customers (regular or close customers) or neighbours. Girls have to fall back on their savings during Ramadan because there are few customers at that time. However, in Baniashanta the girls live a life at risk of natural disasters such as cyclones, tidal surges and floods because their settlement is very close to the Bay of Bengal. The brothel is sometimes called 'death valley' as many girls were washed away during the 1988 cyclone and tidal surge. Sometimes they try to go to the nearby village shelter, but they are not welcomed by the local villagers and for that reason some girls pay to stay in another family's home in the village when disaster threatens. Moreover, they face riverbank erosion every year which forces the relocation of their houses at great cost for such poor women.

Income (Un)certainty Many non-brothel-based sex workers like street and floating girls in Bangladesh lead their everyday lives in miserable conditions. They do not know how much they will earn in a day and some have no income on many days. A few studies with street-based sex workers and their clients in Dhaka found that the girls' income uncertainty is an important barrier for their behaviour change in HIV prevention. Thus, without legal protection under Bangladeshi law, girls and women involved in this profession have to struggle daily for their survival. Although brothel sex workers depend on regular clients for a predictable income, many non-brothel women's daily income depends on their health, place of operation and 'physical appearance' and may be as low as taka 50. Women who are usually prostituting in the street tend to stand on street corners, where they find a large supply of potential clients for whom anonymity is important. As they have no established or fixed places to engage in their work, they are considered floating prostitutes. Most of them are independent, although some rely on pimps for help in finding clients. Bus terminals and stops, railway stations, cinema halls and river banks are the usual locations where the contract is negotiated, from where they go to cheap hotels, buildings under construction, dark parks and, sometimes, the client's own house. The stereotypical picture of their clothing and cosmetic use is sometimes in evidence. Some consciously dress provocatively while others choose to dress neatly. They have to wear very heavy make-up, otherwise they will not attract customers.

Regarding income, many street girls have no fixed rates like the brothel girls. For their survival or consistent livelihood, they are forced to work for as little as 10 taka (equivalent to about eight pence). Occasionally, they find a new client whom they term a '*dhur*' (a gullible person who has just arrived from the countryside), but most of their customers are low-income clients like labourers, transport workers, migrant people or uniformed personnel and they pay as little as they can. Sometimes these girls need to have sex with their regular known customers without being paid as one of their techniques to get them regularly. Most have very poor living conditions think little about the importance of sanitary facilities, showers or hygienic practices. Only a few street-based workers are able to rent a room where they sleep and where they can use sanitary latrines. Many of them have no specific home to sleep in or to take food. They bathe in public ponds, ditches or rivers. Usually they sleep on the street or in bus or railway stations, in parks and take their food in street hotels. In both Jessore and Khulna, I found many street girls who have no access to shelter or hygiene practices. One of them named Subarna in Jessore, who sleeps near the town hall, bathes in the stadium pond and has been living this life for the last 7–8 years with a variable everyday income but fixed expenses. Every day she faces harassment from the police. A few floating girls solicit independently, which provides them with some economic advantage due to not having a broker. They can control their own working rate and working days, but at the same time, the level of violence against such floating girls is high. So, they have to be conscious about risk and they know very well that prostitution is stigmatized. On the other hand, I also talked with some casual sex workers or call girls who rely solely on intermediaries for their customers. If they do not get a call from a broker or a customer, they may have to take out loans to meet their basic needs. Their income is also unpredictable because their customers are very irregular. As a result, their lives are high risk because debt and desperation take over. Customers wanting sex without a condom, and offering above the normal payment, are then difficult to refuse. Such uncertainty puts such girls in a more vulnerable situation with regard to STDs and HIV.

Physical Discomfort Commercial sex workers' health status is a combination of pain and discomfort. They suffer from illnesses and experience real pain from the circumstances often forced upon them. In Bangladesh, a significant number of women do not live in brothels, but engage in commercial sex work under the guise of other professions, such as workers in the garment industry. They are vulnerable to life-threatening disease and violence because of the absence of control over their own bodies and the surrounding environment. The equation is that customers are paying money, but the girls suffer violence, exploitation and ill-health. These in turn may hamper their availability for further work, which initiates what we might call their cycle of vulnerability. Most of the women that I interviewed have faced problems of reproductive tract infections (RTIs) and sexually transmitted infections (STIs) for a long time due to their use of unhygienic clothing during their 'period', along with unprotected sex. The effect of these infections may persist for their whole life. Most poor women, including sex workers, repeatedly use the same clothes to absorb the waste of menstrual discharge and keep it in a dark place without maintaining proper hygiene due to attitudes of stigma in conservative families. Most of the women that I spoke to who are involved in commercial sex work suffer from abdominal pain, urinary problems and vaginal discharges. They often feel weak, with irregular menstruation and always the fear of pregnancy because of unprotected sex. In the brothels, they may have several customers in a day. Their uterus can become swollen, with associated aches and pains, sometimes even loss of sensation in the legs. Moreover, as sex workers, particularly brothel girls, use many skin creams and other cheap cosmetics to enhance their attractiveness, many of them face harmful impacts on their body in the long term.

As sex workers are deprived of their rights, so they are very vulnerable to violence. Regarding residence-based sex workers, some customers misbehave, biting and causing bruising. In almost all cases, customers take alcohol or other stimulating drugs before the start of sex and they become emboldened or aggressive. One girl, Lota told me that '*we cannot predict the behaviour of drunks or bad customers. They like biting and some enjoy beating us, shouting and causing pain and suffering. They stay a long time*'. I found some street girls at Khulna rail station who are very young. One of them, Jorna who is only 13 years old told me that '*I feel pain during intercourse*'. They all need to tolerate these sufferings silently because customers have hired them with money. Most of the girls, particularly street and floating sex workers, have to face two further problems. They told me that customers want to copy what they see in pornographic films. They want to fulfil their desires but do not consider the girl's capacity or her suffering. This may involve bleeding, burning and pain from group sex and anal sex. Such experiences are unpredictable and may involve threats that she will be reported to the local police or '*mastaans*' (muscle men). As many girls cannot ensure condom use, they use female contraceptives to prevent unwanted pregnancies but they become very confused about their proper use. Many of them face numerous difficulties with the various types of birth control methods. As they are poor women, they cannot afford to buy these contraceptives from local markets or clinics, so they use cheaper pills and do not follow a doctor's prescription. They tend to follow their clients' or friends' advice but, after a certain time, when they see that their menstruation becomes irregular or non-consistent or varies in quantity, they become anxious. However, most girls/women soon experience severe side effects, such as headaches, abdominal pain, lethargy, eyesight problems and general weakness. After a while, most workers switch to either a depot injection or a copper-t or intra-uterine contraception device (IUCD) or they develop a preference for condom use rather than pills or injections.

Bonded Girls' 'Subordination to Freedom' There is a chain in the trade of bonded sex workers and they have the least freedom. They have no control over their income and cannot be free from bond by own will. They are required to entertain clients even when they are ill, putting at risk both themselves and their clients. It is widely perceived that many punters often seek young girls or virgins and sometimes offer substantial amounts of money for this, which is one of the factors in the trafficking of '*chemri*' or bonded girls or sex workers. The usual practice of entering the profession is the requirement of an adult woman to swear an affidavit before a first class magistrate or notary public to the effect that she wants to follow the profession of prostitution out of her free will. Such an affidavit is commonly considered as a 'licence'. The majority of prostitutes do not have such a 'licence', particularly those who are engaged in the street, hotel or residence. These 'licensed' young sex workers need to earn a lot of money in order to pay off debts to their brokers or owners or *sardarnis*. They have to hand over their earnings because the leaders create a captive situation in the brothel. In Khulna and Jessore, I observed that some minors under 18 years old are involved in sex work but they say that they have a 'licence' where it is written that they are 18. According to the Bangladesh constitution, anyone over 18 has the right to choose their own profession. But this system of affidavits is loose and brokers exploit it. When a new girl comes into the brothel, someone will buy her and she becomes a *chemri* under her *sardarni*. Basically, *chemris* have to earn for their *sardarni* who do not need to work. When *chemris* cannot earn according to their desire, then they beat the girls and mistreat them. The system is that all the earnings of a *chemri* will go to her *sardarni* and if she (the *chemri*) is given any '*baksis*' or extra tips, sometimes the *sardarni* will take that as well. So, they always live a sorrowful life. Regarding their subordinate girls' freedom, the *sardarnis* have a saying that '*we bought the cow from the market, so she must produce milk*'. This bondage lasts from a few months to a few years. It depends on the *sardarni's* behaviour and mentality.

Brothel Girls' Common Dream In Bangladeshi society, most women typically want a life with hope in which there is a home, husband and children. Although sex workers have their own dream of leaving the brothel, they are not always clear how this may be achieved, and their dreams are often clouded by fear. The brothel-based NGO officials that I spoke to confirmed that almost all girls have a dream to leave but cannot because their lives are always in crisis. For example, many times girls take loans with high interest from their house mates, colleagues or *sardarnis* in order to contribute to their parents' welfare or some other family purpose. In this way, they become tied to their leaders or *sardarnis* in financial bondage. They cannot save money from their daily income as the amount is not enough. This cycle of crisis leads to frustration. They are often confused about their real destination due to their uncertain life. When I asked a question about dreams for the future, many workers said that they had none. After some time, they relaxed and tried to think about their future life target. One girl, Parul told me '*what I will do in future? I have no dreams. Everything has gone from my life!* [Then after some thought] ...*Will I do this work my whole life? No, I will go to my village and look after my baby, bring her up and spend money on her education*'.

Many sex workers have a common dream to marry a 'good' man or be rescued from the brothel by a good man. According to their idea, a good man would not give them any trouble, would not beat them and use vulgar words and would not have an exploiting mentality. Many girls try to save money for a dowry and leave it with a trusted person, but theft is common under such circumstances. As the girls usually cannot get their desired 'good man' in life or be cheated, so some have recently been trying to get involved in self-employment training provided by different NGOs, for example, candle making, or handicrafts. During the course of my field work, one day I found some Fultola brothel sex workers at an NGO office being trained in candle making. They told me that they earned lots of money in their life but right now they needed training as they know that they cannot do this work forever. I asked them about their future destinies and they responded:

> (Shefali) 'I have a desire to make a family, have a husband and children; I will buy some agricultural land with my savings'.

> (Nipa) 'I have a dream that when I will leave the brothel, I will give my savings to a good young man to develop a business and will marry him and enjoy family life in peace and happiness'.

> (Kobita) 'I will develop a co-operative for handcrafts with some other girls for self employment. The whole day we will work hard and in the evening we will sell our products and eat and sleep. We need happiness.'

Most of the sex workers strongly believe that they are cursed into the life of a prostitute. Many of them think that it is a way of life and a way to make money. People who hate them for their profession do not understand the hardships they are forced to endure. They also believe that as they are poor, so they have to adjust to this life and they have to live with laughter, tears and sorrows. They also believe that if anybody is not fated to have happiness she cannot get it even by paying money. However, in the brothel, sex workers are isolated and cannot use government facilities. No one seems to be taking responsibility for their rights like hygiene and security. The politicians blame police inaction and the police cite the problem of the house owners or *sardarnis*. The problem does not seem to be a priority for anyone.

Brothel Girls' 'Love Trap' Brothel girls consider themselves to be 'fond of love'. Almost all of them are being exploited in the brothel by their so-called lover or '*dada babu*' or '*babu*' which seems to be a 'love trap' for them. In many brothel, it is very common that women have *babu* or steady lovers and it is believed that they do not use condoms in these private relationships. These are former customers with whom the girls share their dreams. But the dream of rescue and marriage, without exception, never comes true in reality because most of the *dada babus* are unemployed and they stay in the brothel to solve their own problems. I was told that it is common for the *dada babu* to run off with money and other belongings, but he may soon be replaced by another.

During an FGD with Jessore brothel girls, they told me that they all want to love someone but that 'brothel love is fake'. If they leave to live with a *babu,* then most return to the brothel after a while after their money runs out. Here are two stories of a brothel sex worker's relationships with a *babu* or *dada babu*. The first tells of the girl's hope; the other about exploitation.

Future hope—Sapla is a sex worker at Baniashanta brothel:

> 'I have a dada babu who has no relation with any other girl. I love him because he looks very nice. Sometimes my dada babu stays with me during the night. Recently he has not been allowing me to take any customers. I have many dreams about him. I don't believe that my dada babu will act like the others. He will not steal my money and belongings. You

cannot compare my dada babu with others because mine is exceptional. His mother and the whole family know about me, that I am staying here to make a business. He spends money on me because he considers me as his wife. If my dada babu notices that I am talking and smiling with other men he becomes very angry; then I know that he likes me. I think that life is a combination of good and bad. I believe that I will go to my husband's house very shortly'.

Exploited victim—Santa a sex worker at Baniashanta brothel had a *dada babu*.

'One of my regular customers forced me to take him as my dada babu when I arrived in this brothel. If I hadn't agreed he would have arranged an 'accident'. He was a mastaan in the area, not a good type of man. He was involved in burglary in the forest area of the sundarbans. He stayed with me and lived on my income. There was a lot of trouble and he abused and swore at me. When I left him, I lost all my savings and ornaments. Now I don't trust anybody'.

Babus are very influential men in the brothel. Although some are abusive and exploitative, most of the women value their relationships with these men. Most of the *babus* have wives and children outside the brothel and some are *babus* to women in other brothels as well. It is highly likely that these women are at risk from the multiple sex partnerships of their *babus*.

Non-brothel Girls' 'Temporary Love' Women sex workers have been blamed for the spread of HIV but, as frequently pointed out, sex work exists only because of the demand from men. The casual sex workers I spoke to told me of their techniques of sustaining a livelihood. When a girl finds a friend-cum-customer, they maintain the relation for few visits and tell their 'miseries' and take the opportunity to borrow money from the man. Such relationships turn into sex after some time. Usually a girl will be loyal to her friend and not look for other customers. But after some time, ranging from a few weeks to few months, this kind of temporary love or friendship stops due to the man's lack of commitment. He may realize that she is a call girl or lose the taste for sex, and may break the relationship by mistreatment or making an accusation. The girl will then look for another potential man who will give her money in exchange of 'temporary love'. I found several call girls in Jessore and Khulna who described the above situation.

Many hotel girls are mentally depressed due to their financial position. In some hotels, girls stay for a few days and perform sex acts on a contract basis. They cannot protest about any inflated bills for fear of facing physical and mental abuse. They put up with it because they have debts in the village which need to be paid. I found one hotel and residence-based sex worker, Lima who had physical problems for a long time due to her own negligence of her body but she did not feel able to see a doctor. Others become pregnant by customers who refuse to use a condom and then insist on an abortion. I spoke to Sathi who had recently been through this process. She now feels very frustrated. Others turn to alcohol or cutting themselves to release their anger. One girl, Ripa told me that:

'When the girls feel sad, they take sedatives. They want to forget their sadness. We are deprived from many things, such as family, relatives, and a husband. When I want to cross the road, people comment on my identity. We cannot go to our parents or talk to relatives'.

There are also many frustrating situations prevailing in the sex worker's life. They openly complain about exploitation and cheating by various powerful groups such as the police. It is commonly reported that most of their earnings go to exploitative groups, such as pimps, madams, muscle men or corrupt officials. Some sex workers have also experienced difficulties including abuse from relatives or previous lovers resulting in a certain level of depression and other problems. Recently, however, there have been some improvements for sex workers through NGOs' provision of training on rights. These capacity-building lessons are based on avoiding harassment, rights to proper health services and the empowerment of the girls.

2.3 Drug User's Daily Vulnerability

Drug abuse is defined as taking a drug to such an extent that it greatly increases the danger or impairs the ability of an individual to function or adequately cope with his or her circumstances (Irwin, 1973). Experts in this field agree that although no factors actually cause substance abuse, there are several that contribute to its gradual development (Lex, 2000). Generally, a person with serious life stresses, or major depression, or a tendency to curiosity and experimentation, may be at increased risk, but is not fated inevitably to become a substance abuser. These drugs may relieve pain, induce sleep and stop diarrhoea, but tolerance and dependence sometimes develop at therapeutic dosages. Psychoactive drugs are found in three major groups: depressants, stimulants and hallucinogens (Whelan, 2004). The use of drugs and spread of drug cultures are reflective of these everyday tendencies towards opposition to 'rationalistic' and 'regulation', and the embrace of indulgence in dionysiac activities and the 'carnivalesque' (South, 1999). Substance abusers exhibit complex problems. In the past years, multiple substance use has become common and a lifelong disorder—a 'chronic relapsing disease' (Lex, 2000). As substance abuse is a relapsing disease, there is no cure but there are several paths to recovery (O'Brien & McLellan, 1996). The health consequences of illicit psychoactive substance use are numerous and diverse and are related not only to the direct action of the drugs themselves, but also to the mode of administration and associated lifestyle factors (Bungay, Johnson, Varcoe, & Boyd, 2010; Lex, 2000; Sutherland, 2004). Kellehear and Cvetkovski (2004) characterize the drug user as respectively, morally bereft, ill, or a social victim. The relationship between drugs use and crime may be more profound than simple causality (Harris, 2007). Victimization is the understanding of the processes associated with the impact of crime (Walklate, 2007). All over the world, drug users have been characterized as leading 'unstable' and 'chaotic' lives (Singh, Sivek, Wagener, Hong Nguyen, & Yu, 1996; Williams, Bowen, Ross, Freeman, & Elwood, 2000) as indicated by the lack of permanent housing and/or employment, involvement with the criminal justice system, absence of ongoing relationships and/or reliance on public entitlements (Rhodes, 2009). Evidence shows that some of the influx to injection drugs is due to drug users switching to injecting after taking other drugs for 4–10 years (NASP, 2003). Injecting drug use generally involves greater health risks than non-injecting use (Sutherland, 2004).

Drug abuse is a significant problem in Bangladesh, where the majority of addicts are thought to use the mode of inhalation. Although no scientifically rigorous surveys have been carried out, it is estimated that there may be as many as 4.6 million Bangladeshis who abuse drugs of which 91% are adolescents/youths (Sharma, Rahman, Kelly, & O'Day, 2006). While the vast majority of drug users are young male (Kabir, Goh, Kamal, Khan, & Kazembe, 2013) and heroin smokers, there are at least 20,000–40,000 injecting drug users in Bangladesh. The most common injected drug in Bangladesh is buprenorphine that produces opiate-like euphoric effects and dependence (Koester, Rahman, Kelly, & O'Day, 2005). Mahbubur Rahman, Zaman, Sekimoto, and Fukui (2000) also found some HIV related risk behaviours particularly needle sharing among drug users are common in Bangladesh. To estimate the prevalence of hepatitis B and C, and HIV infections among drug addicts in Dhaka, Shirin, Ahmed, Iqbal, Islam, and Islam (2000) found the HCV infection was significantly higher among the IVDUs who are associated with the sharing of needles and a longer duration of injecting drugs. On the other side, the sero-prevalence of HBV infection was significantly higher among the IVDU and non-IVDUs who had a history of extramarital or premarital sex. Moreover, there is general agreement that socially and economically marginalized groups are more prone to drug use and its more serious form, problem drug use (see also Islam et al., 2015). The abuse of drugs in Bangladesh is closely associated with other social crimes such as theft and robbery (Hossain, 2005). Most drug users supplement their income by pick pocketing, thieving or selling their blood (Knight, 2006). However, Rahman and Zaman (2005) found an intertwined relationship between drug use and unsafe sexual behaviour among male drug users in Bangladesh (also in Hassan, 2005).

Other than commercial sex and needle sharing, many street-based drug users take other risks, such as sleeping on the streets or in other dangerous places where accidents occur. They take risks like theft to raise money for drugs, and face a beating if they are caught. Moreover, a good number of drug users have a tendency to switch from one drug addiction to another, particularly from heroin to injecting drugs or vice versa. This mainly depends on cost and availability of alternatives and the withdrawal symptoms suffered in switching. This section aims to focus on the drug addicts everyday suffering from withdrawal symptoms and the many negative impacts on the body by drug abuse and drug switching. However, their desperation for drug money and its consequences will also be addressed. Moreover, along with vulnerability issues, how the addicts follow a life cycle of relapse and recovery. In addition, how the 'surrounding environment' of the addict plays a role in 'polluting' the addicts.

Addiction Act as 'Vampire' Many behavioural scientists termed the drug addiction as a human-made illness which has its roots in socio-environmental condition. In most cases, the opiates are obtained easily and without cost for their initial doses. Bangladeshi addicts take some simple drugs like cannabis as their 'starter'; eventually many enter the world of injectable drugs and finally graduate to deadly drugs such as heroin. Sometimes heroin addicts switch over to needle use but they come back again to inhaling heroin. Most of the drug addicts, particularly the heroin users that I spoke to in Jessore and Khulna, got involved with addiction mainly from curiosity about something new or by keeping bad company. One of the respondents in Jessore named Kamrul told me:

> 'Drug addiction including heroin and injecting drugs comes from curiosity, family quarrels, bad company, the failure of love affairs, mental stress, unemployment, sadness, all are responsible. When you can't see a way out, it's like medicine'.

There is a common saying about heroin addiction that 'you may want to leave me (heroin), but I will not leave you'. Drug addiction, particularly heroin addiction, is like 'vampire behaviour'. In the movie we see that when a vampire bites anyone who is resisting, the victim becomes in turn a vampire themselves. So, it is with heroin addiction. Dealers will deliberately encourage potential addicts in order to create easy business. After a time, the addict will become a dealer and so on, in a chain. Many times during my field work, I heard of innocent people who had been lost to drug addiction. An example is the young protester who tried to rid his area of addicts. Then all of the local peddlers tried to involve him in addiction by providing free cigarettes which contained a small quantity of heroin. When he asked for more, he was introduced to a dealer to buy it and learned how to take it. When he managed to come to a treatment centre, then again he became determined to evict these people from his area. Although they have different reasons for their addictions, all of them suffer from the same physical problems. Addicts told me that although when they take drugs they feel 'the best', actually they become 'the beast'.

Everyday Sufferings with Withdrawal Symptoms Many works show that psychoactive drugs like heroin and morphine have the ability to banish all aches and irritation but the experience of withdrawal is often considered as the most vital aspect of drug addiction. Withdrawal symptoms generally start from 12 to 24 h and it progresses with additional signs and symptoms. There are some physical symptoms of withdrawal syndrome such as aching limbs, a runny nose and shaking. Mental symptoms, such as disturbed sleep and anxiety can last much longer. There are some well-observed pathways to heroin addiction like sniffing (snorting), inhaling heroine vapour (chasing the dragon), smoking, skin-popping or intravenous injection. In Bangladesh, chasing the dragon is popular, along with injecting. To inhale heroin, a dose is placed inside a strip of foil paper, such as that found inside cigarette packets, and a flame is run underneath it. The heroin heats up and turns to vapour, which is then sucked up into the nostrils via a tube. Those drug users who prefer to take morphine through injection target the large veins in the forearms, thighs, buttocks and stomach.

Drug users suffer physical and psychological symptoms if they cannot obtain drugs. Most addicts have to take them on a daily basis and it becomes a way of life for those addicted. When an addict's withdrawal symptoms or '*bera*' appears, they go to collect or raise money for buying drugs, and after taking drugs they become 'calm'. Without taking drugs they avoid normal food and find no taste in it. Some users take drugs three times a day, rather like breakfast, lunch and dinner. They take heroin as a stimulant, and then sedatives for sleeping. In Jessore, one heroin user, Shapon told me his story:

> 'If you don't take heroin or other drugs timely, then a hundred types of sufferings will appear in the body, like fever, cold symptoms, body pain, coughs, sneezing, headache, diarrhoea, insomnia, burning sensations and itching in the whole body, drowsiness and red eyes, stomach ache, a weak heart, emphysema, bleeding gums, and many other physical problems. After taking heroin we go to another world, where there is no tension, no thought for this world, and everything is fine'.

When a person starts heroin inhalation or other drug taking, s/he experiences some feelings which normally continue for a few weeks, but when this stage finishes, they need to take more drugs to remove the bad feelings or get rid of withdrawal symptoms. Some heroin smokers believe that the heroin presently available is not like previous batches in purity. They say that it has been contaminated by dishonest dealers in order to increase their profit. Due to contamination, they say they now need to take many times the previous quantity to support their body, instead of just one dose.

Negative Impacts of Addiction on the Body Evidence shows that opiate drugs like heroin have the capacity to alter mood and behaviour. These drugs can produce two major groups of harm: first, toxicity and second, dependence. This opiate drug is called as most risky substance of the world. Indeed, heroin is a powerful and stronger narcotic than morphine which suppresses both respiratory and cardiovascular activity along with strong painkiller effects. The main medical disorders listed as 'complications of heroin dependence' are depression, blood vessel infection, nerve damage, pneumonia, fungal infection of the heart and brain damage. Once in the bloodstream, opioids are distributed throughout the body, with accumulations in the kidneys, lungs, liver, spleen, digestive tract and muscles, as well as the brain. Depressants, like sedative/hypnotics, alcohol and heroin are drugs that slow down the activity of the brain. Depending on the dose, these drugs produce relaxation or drowsiness, emotional depression or loss of consciousness. In particular, heroin is especially likely to cause coma by depressing an individual's breathing.

I found some drug users who had bronchitis and severe coughs. It is usual for drug users to have multiple physical problems due to serious and continuous involvement in addiction, and most of them face negative impacts on their bodies in the long run. Among the commonest complaints are lacks of strength, respiratory problems, weight loss due to lack of appetite, kidney damage and scabies. In Jessore rail station, I found one drug addict, Giash who had been suffering gangrene in his leg. He has no family or relatives and so he has to manage on his own to get drug money. He shows his gangrene infection to passers-by to get sympathy and begs for money in the name of treatment. He does not take any medicine and instead spends the money on drugs. However, I did find one addict in an FGD at Khulna named Raju who managed his drug money in legal ways. He thinks that drugs usually 'eat' the addicts, but in his case drugs are not able to eat him. He said '*I can work because*

I take the drug and food simultaneously on time, I bathe regularly and, because I have no tension like others for drug money, so it doesn't hamper my activities'.

One reason that addicts feel weak and anaemic is that selling their own blood is their last option to raise drug money. I found a former rickshaw puller and heroin addict who had recently sold his blood and was left feeling so weak that he could not walk and was in fear of his life. But I also found many drug users cum professional blood donors who sell their blood frequently, maybe twice a week and believe that if they take a vitamin tablet, their blood will recover and they will be fit again. From my observation in Bangladesh, drug users feel well at first but after a few months to a few years they become depressed. Their level of hygiene and intelligence goes down, and this starts many physical problems. Most drug users' prevalent health problems are malnutrition, protein and vitamin deficiency, and anaemia. In general, they neglect themselves because many of them lose interest in their appearance and also in social conventions. As they cannot afford a proper diet, weight loss is a common consequence of heroin misuse. Skin problems like itching are also common due to a combination of poor nutrition and infrequent bathing. These can increase discomfort and may cause general ill-health. Indeed, injecting drug users commonly face abscesses and other skin problems which are usually the result of a failure to inject the substance completely into the vein. However, many addicts seem to face bronchitis or asthmatic problems, perhaps due to living on the streets as well as cigarette smoking. Moreover, many addicts are susceptible to respiratory diseases, infections of the urinary tract and kidney problems.

Drug Switching and Needle Sharing Heroin users are at high risk of multidrug abuse and dependence. They may find their tranquilizer supply interrupted and then use other drugs. Intravenous drug use generally involves high health risks than non-injecting use. Evidence shows that many drug users change their drug, for example, from heroin to injecting drugs within few years. The most common injected drug in Bangladesh is buprenorphine that produces opiate-like euphoric effects and dependence. To estimate the prevalence of hepatitis B and C, and HIV infections among drug addicts in Dhaka, it is found that HCV infection was significantly higher among the IVDUs who are associated with the sharing of needles and a longer duration of injecting drugs. On the other side, the sero-prevalence of HBV infection was significantly higher among the IVDU and non-IVDUs who had a history of extra-marital or premarital sex. It is also found that HIV related risk behaviours, particularly needle sharing among the drug users are common in Bangladesh. There are many reasons for 'switching-over' from one drug to another. For example, when there is a heroin shortage in the market, its price increases; as a result, poor addicts switch over to injecting drugs, which are comparatively cheap. This is a recent phenomenon because, until a few years ago, these were pioneer opiates until heroin came onto the illegal drug market and addicts jumped into it from injectable drugs due to its low price. For example, one ampoule of pethidine costs more than 100 taka, whereas heroin comes at 50 taka (a small wrap) a fix. Injecting drug users (IDUs) usually share their needles with their close friends. In that case, place, time and money are mainly responsible for needle sharing among the addicts. Addicts cannot wait to acquire a clean needle, so they share. One IVDU named Rakib told me that '*sometimes we share the same injection syringe; ampoule sharing also happens. We use the same syringe among our friends when the shop closes, and we cannot delay*'.

One serious drug addict in Khulna, Rafiq told me of a rule that, if any of his friends manage to get any drugs, there is an expectation that he will share. He also said that they use the syringe to take some blood from their own bodies, and then hand over to others to inject, so that they can get a high. It is a significant risk factor in the spread of HIV, hepatitis B and C and other blood-borne infections because of the common practice of sharing needles. Repeated injections can cause vein collapse or loss. Drug users inject directly into their veins rather than their muscles because of the faster reaction. But they may subsequently face the problem of finding a vein because of their frequent use of needles. One showed me his vein shrinkage and I could not see any veins in either arm. He was concerned because he had heard that his vein problem might prevent him from working. As a result, some addicts inject a finger or penis due to the disappearance of veins in other parts of the body, or they inject drugs into muscles rather than veins. These injecting drug users suffer abbesses and intermittent bleeding. Moreover, they lose sensation in parts of the body.

Stealing and Facing Retribution There is a close relationship between drug addiction and crime. Most drug users lack the funds to purchase sufficient opiates on a regular basis which ultimately push them to commit crimes in order to obtain the substance. And this is how drug users become victimized to the society. Many addicts have to make choices about selling their favourite or family belongings like jewellery or wrist watches to purchase heroin or other drugs. Heroin addicts and other drug dependents manage their drug money on a day to day basis. Study shows that the average per person cost of drugs was US$707–1135 per year, which is much higher than the per capita income of Bangladeshi people. Most drug users supplement their income by pick pocketing, thieving or selling their blood even. In other words, the abuse of drugs in Bangladesh is closely associated with many social crimes such as theft and robbery.

Drug users follow many strategies, ranging from begging to stealing. Some collect rubbish (old papers, aluminium jars and animal bones) from residential areas and sell it. Some beg by showing any infection they have on their body, or by presenting themselves as destitute, or any other tactic to collect money or rice. Sometimes people identify them as heroin addicts and refuse to give. When they cannot manage to acquire money in a legal way (mainly from family, rubbish collection or begging), then they are prepared to steal. If they are caught, they may be beaten up and sustain serious injuries. Often addicts carry a blade in their mouth or in their pocket when they go to steal. If faced by an angry mob, the addict cuts his hand and people are scared off by seeing the blood. He may even get some sympathy. Saju told me about his technique for collecting drug money:

> 'When I go to any house, and if there's no-one there I take anything I can carry. If I'm caught then they will beat me and I have many fractures in the body. Look, many joints of my body are already broken due to beatings. I bring stolen goods to certain shops in the railway station that are willing to buy, and the money goes on heroin. Sometimes I beg by showing my broken leg or hands. Some people can understand this and if I get, say, 3 kg of rice I sell it and use the cash for drugs'.

The evidence indicates that drug abuse is associated with an increasing number of people who commit crimes and, as a result, many addicts suffer from physical assault. Some experts claim that drug use is a mental illness to some extent and that there is a close relationship between mental attraction and drug use along with physical symptoms. As many drug users are unable to mix in mainstream society, they always live in a sense of depression, which makes them frustrated and anxious. In most cases, due to substance use, many drug users fall into a pattern of victimization.

Drug Users' Sexual 'Pleasure' Addicts develop different attitudes during addiction. There is an 'invisible' relationship between drug use and unsafe sexual behaviour among male drug users in Bangladesh. When people, particularly married men, indulge in heroin, one reason is that they have heard that it prolongs sexual 'pleasure' due to long intercourse. Some reported that their wives accused them of a kind of sexual torture as a result, and, rather than risk discovery of their drug addiction, they find release elsewhere for what they claim to be a boosted 'sex power'. The new sex partner may be the source of or a victim of STDs. I found one injecting drug user, Rassel who is married but very interested in outside sex. He mixes with family girls (those who are not treated as bad girls), hotel girls and sometimes brothel girls on a regular basis to fulfil his sexual urges. When he takes drugs he becomes crazy for sex, and then he forgets his wife because he thinks that he can enjoy himself with her anytime and she is '*dal-vat*' (familiar) to him, and so he wants to 'taste the flavour of other dishes'. He told me another important reason:

> 'My wife is not able to take my pressure for long intercourse. She becomes fed up, besides, other than sex, I need to do many things on the girl's body like biting and punching heavily, which behaviour would not be tolerated or allowed by my wife. From the outside sex, I have some scabies and itching problem in my penis because most of the times I don't use a condom'.

I found an unmarried heroin and injecting drug user, Didar who has a sex addiction. He told me that some of his friends inspired him to enjoy sex after taking drugs. From then onwards he became keen on girls as a recreation along with the drug addiction. He has visited almost every hotel in Khulna and all the nearest brothels. Besides, he 'uses' *family girls* for his sex addiction. Regarding condom use, some girls insist on condom use but he knows of hotel girls who will allow him to have sex without in which case he sometimes pays extra. Like Didar, many drug users' condom use with sex partners depends on the type of relationship. Recently, he has been facing serious physical problems in the genital area but feels unable to consult a doctor.

Helpless Addicts' Relapse and Fail to 'Re-live' The causes of relapse can be divided into two indications named individual and socio-environmental. The first cause for relapse includes physical and psychological sensations. The second type includes socio-environmental influences such as drug smell, availability of drug selling points or people with drug use. In Bangladesh, most of the drug patients in the treatment centre feel that they will eventually recover. Their mental strength seems remarkable, but when they come out from the treatment centre they return to the context of addiction in which they can save themselves for a few weeks to a few months at best. Heroin addicts are attracted by the smell of the smoke and it is easy to fall into the same trap of relapse. There are many factors causing relapse, like mixing with addict friends again, or over-confidence about staying clean. I met one drug addict, Kabir who has been taking drugs for a long time. He asked me a vital question about the value of drug treatment:

> 'People can get treatment from NGOs, but they return again to drugs because this environment is polluting. If I don't change the place it will not be possible to change my addiction. If I take the treatment in the centre, and stay for 15 days, the government will spend 15,000 taka on me, but this money is a waste if I start taking drugs again. So, what is the value of the money?'

When drug addicts take treatment, they may reflect on their life of addiction. I talked with one addict in a rehabilitation centre who was addicted to injecting drugs and now felt that he had made his family suffer. He also thinks that he had no sense to judge what was right and wrong. Many drug addicts want to go back to mainstream life but they cannot tackle the suffering, and control their own minds. Many addicts' families also try to help them return to normal life. I interviewed one addict, Shapon who wants to help himself:

> 'Many times I think that I will surrender my life under a truck, commit suicide, because I have no income. If I am able to manage one meal, the next one may be two days away, so why I will keep this life? Sometimes I feel I want to get back to normal. I can remember my school life'.

I found a boy named Islam in a drug rehabilitation centre in Jessore during a participant observation session. Islam does not know what will happen in his life after finishing the treatment course in the centre. He actually cannot confirm that he will be able to leave the addiction or not, or when he will leave the rehab centre. However, I met with some drug patients in the same treatment centre who thought that if they lost the opportunity, they would go mad and destroy themselves. Drug addicts believe that they can leave the drugs if the environment helps them. At the same time, family support is also important for leaving the addiction. Recovering addicts have a high social pressure to prove themselves to be drug free. The ready availability of drugs is an important issue in compromising the recovery of addicts.

Many addicts eventually try to stop using heroin on their own or at least to cut back on their dosage as a part of self-control. Some realize that they have become alarmingly thin and unhealthy looking, that they are an economic burden for their family, that they could no longer participate in communal activities of the neighbourhood or village or that their social reputation has declined. However, most of the time, particularly after taking the drug, they come to believe that they are harming themselves by their addictive behaviour. Then they decide that they would be better off if they were not addicted. But the addict cannot keep to these wishes for long, because in the meantime, withdrawal symptoms appear. As relapse is fairly common in recovery from substance addiction, it should be incorporated into treatment planning.

2.4 Sorrows and Anger of PLWH

Although a large body of literature currently exists on HIV/AIDS in general, rather less research has explicitly considered the geographical dimensions of living with HIV/AIDS. In other words, geographical studies focusing upon the experience of people with HIV/AIDS are limited. Geographers have begun to explore the nature and extent of the impact of HIV/AIDS on health services, and on diverse aspects of the daily lives of people living with HIV and AIDS (for example, Del Casino, 2001; Takahashi & Smutny, 2001; Young & Ansell, 2003). This research has provided a further illustration of the way in which health status is the product of more than health care alone, a theme which has received considerable attention in discussions about the evolving structure of 'medical geography' (Kearns, 1993, 1994; Dorn & Laws, 1994; Hayes, Foster, & Foster, 1994). Some studies of the social construction of HIV/AIDS also include exploration of the stigmatization of HIV/AIDS, a central component of the developing literature, a key concern of this book and discussed in further detail later in this chapter.

The impact of HIV/AIDS influences every aspects of the life of PLWH (Elmore, 2006). Most people who suffer from AIDS may have been infected with the virus for a long time (often several years) before showing signs of its development (Haynes, Pantaleo, & Fanci, 1996). Moreover, not everyone who is infected with HIV reacts in the same way. While some PLWH individuals succumb to AIDS within a few years, others may remain healthy for a decade or more (Thompson, 1996). Adaptation to chronic illness that encompasses the short- and long-term stresses and strains is presented by living with a chronic condition (Luginaah, 2008; Revenson, 2001). Living with HIV/AIDS on a daily basis is a particularly complex and difficult task (Wilton, 1996) as it lacks family and health support from family and friends (Elmore, 2006). It is hardly surprising that a wide range of adverse emotional reactions like suicide risk (Campbell, 1995) are regularly reported in PLWH. Individual reactions to PLWH are likely to be the more painful and severe (Gagnon, Merry, Bocking, Rosenberg, & Oxman-Martinez, 2010) because it is usually suspected that the disease is spread by sexual encounters. HIV has been linked to a wide range of neuropsychiatric syndromes including depression and anxiety. Perhaps the most pervasive feelings voiced by AIDS patients are anxieties and anger over the uncertainty surrounding their illness and treatment including family ignorance (Gilbert & Walker, 2009). Their questions about treatment and lifestyle modifications following acute illness are generally met with incomplete and unsatisfactory answers (Collins et al., 2016). As in other developing countries, PLWH in Bangladesh have to face many unexpected matters which make them very frustrated and hopeless in their daily life. This section will look at these issues of how PLWH express their sorrows in their everyday life and how they try to cope.

Insulting Language In Bangladesh, many PLWH feel upset at the insults that come their way, when they themselves do not feel at blame, especially women who have been infected by their husbands. Some NGOs have been presenting PLWH in different forums as speakers to create a mass awareness about the HIV issue among ordinary people. Yet in discussion meetings on awareness, PLWH often face unbearable insults. In Khulna, I met with some PLWH who are working as NGO staff and appearing in different seminars despite their fear of the reaction. The main objective of these seminars is to share their suffering and raising awareness. But, instead of offering sympathy, the audiences tend to be very negative. One female PLWH, Diba who has been to many forums told me that it is not uncommon for people to say that all 'positives' should be killed so that they cannot spread the virus. There are also sneers about the obvious inadequacy of their prayers; or how long they will have to wait for death, and how are they passing the time? Moreover, some would prefer to die because of the misery and money crises they face. One PLWH, Keya, told me that she is alone in the world and cannot express her sorrows to anyone. She believes that she is the most hopeless girl in the world. She has travelled to many places for a livelihood, like a floating object, but she cannot stay anywhere for long and never finds any peace of mind.

The Dilemma of Marriage The growing and continuing challenge of defeating AIDS has become more concentrated among women in the developing world, particularly countries in south Asia, where increasing rates of infection have shown women innocently succumbing to infections. HIV infection puts extraordinary stresses on people's lives. Most of these stresses are unusual and people are unsure how to handle them. During my field work at Khulna, I found a few husbandless PLWH women (mainly abandoned) who want to get married. Some confessed that they had had offers of marriage but they do not have the courage to proceed because if they get married to a 'negative' man, the NGOs would stop their medical and nutritional support, and then would the husband be able to support them and provide for their needs? There are examples of marriages between 'positives' and 'negatives' where the 'negative' husband had not been infected. So many of them have the wish but they also have fears, sadness and a serious dilemma.

Feelings About Hopelessness Many PLWH are living their lives not only with physical suffering but also with psychological stress. Among the PLWH in Khulna, I met a woman who had married a Bangladeshi man when he was in Malaysia as a migrant worker. Although her family were cautious about her love affair with a 'foreigner' because of the long distance from Malaysia to Bangladesh, she was very serious about the relationship. Now she thinks that she made a mistake to take him as a lover because he did not mention anything about his HIV status before marriage, although he knew it. She only learned about it when she was brought to Khulna Medical College Hospital by her husband and felt an intolerable shock. Despite that, she considered it to be her fate until she realized that her husband was involved with other women, and then she felt hopeless and considered his behaviour to be a betrayal.

Physical Suffering In Bangladesh, PLWH commonly and continuously face diarrhoea problems, mouth ulcers, fever, colds, sneezing, coughs, tuberculosis (TB), herpes and skin diseases as 'opportunistic infections'. One PLWH named Afsar recalled how he had suddenly noticed that his whole body, excluding only his eyes, was covered with a chicken pox type skin problem. There was a bad smell because of the pus and his body swelled up. He went to hospital and stayed there for three weeks. After blood tests they discovered that he had HIV. After a few weeks, he got back his normal skin. A PLWH widow, Shanti, who was an agricultural labourer, worked at cutting the crops and found that if she had any injury, it would not recover. She also lost her taste because of the many ulcers in her mouth. She became black and very thin, with many abscesses, diarrhoea, stomach pain, headaches and frequent fever. She visited Khulna Medical College Hospital twice but they could not trace her problem. Some doctors diagnosed ulcers, and others mentioned cancer. Finally, after blood tests doctors identified her as PLWH.

Expected Longevity In Bangladesh, many newly identified PLWH do not know how long they have been suffering from the disease. Some have died within two years but others are still living who were diagnosed in 1997. The longevity of PLWH depends on many issues. Experts and NGO professionals believe that this longevity depends on medical and nutritional support. Those who can afford good food and nutritional support, along with medicine, have extended lives, mainly where there is opportunistic infection along with ARV (anti-retroviral therapy). Many of the PLWH themselves think that their life will not be sustained for long and I found very few PLWH who wish to survive as long as a normal person.

Devotion to God Despite anger and depression, almost all PLWH, irrespective of gender and different religions are very devoted to their own god. The vast majority of the PLWH participants described religious faith as a major source of support in coping with their sorrows and anger. Most of them believe that religious devotion to god can provide relief from the physical pain and suffering from HIV and different opportunistic infections. They also believe that devotion to god also keeps them away from mental suffering due to having HIV status. They think that religion will purify their mind and help them to be devoted to the god. I found many PLWH who believe that their life is for god and they have no pleasure without calling on god. Out of many, one of the female Hindu participants, Shanti, a former housewife, believes that her HIV disease was, in a sense, given by god. Although she has not done anything wrong, her husband did and she believes that when she got married she became part of his sin and virtue. As her husband was sinful, so she got a share

(HIV) from god. Despite religious practice, she feels sad for her physical suffering due to her innocence. Some PLWH believe that god has given them the disease and that god will remove it. So, some of them are not scared. One PLWH, Jhorna, a former sex worker, told me that she believes that Allah has given her the disease so that he will rectify her life from bad to good. On the other hand, a few believers are angry with god for visiting disease on innocents.

Scaring HIV Identification Most people feel shocked and disorientated when they discover they have AIDS or are infected with the virus which causes it. They may refuse to accept the diagnosis and become angry. Alternatively, they may react by blaming themselves and feel depressed. Almost everyone feels anxious and scared. Anxiety is something we all experience at different times in our lives. In this context, however, the anxiety felt is likely to be far more severe and longer-lasting. In Bangladeshi society, the lack of knowledge about HIV is such that many 'positives', when they are first diagnosed, are not aware of the implications. They need counselling, but unfortunately often their first experience is to find themselves the object of curiosity. There seems to be no guarantee of patient confidentiality, whether in government or private medicine and the news leaks out quickly. One man, Dulal, told me about taking his son, who had diarrhoea and anaemia, to a clinic. The boy needed a transfusion but there was no O+ blood available, so he volunteered to donate his own blood. He was tested and was shocked to be identified as PLWH. He then brought his wife and all of his sons to the medical centre and, tragically, all tested positive. This information was divulged by the lab attendants and hospital officials and soon the whole town knew. They became objects of morbid interest. According to him, more than 2000 people including journalists gathered in the clinic and were looking at him strangely like an animal from the *Sundarbans* (mangrove forest of Bangladesh). Then he felt frightened and was crying because everybody was saying that he will be killed or likely to be burned. He became the 'talk of the town' in Khulna as well as in other parts of the country.

PLWH people's reactions differ widely, but nearly everyone shares to some extent feelings of anger, depression, fear and guilt. One reason for anger is the unfairness of the situation. Depression is one of the most painful feelings a person can have. For almost all PLWH, depression is both a common and an understandable reaction. Most people become depressed when they find out that they have AIDS. AIDS makes them feel that they no longer have any control over their lives which may contribute to a feeling of loss of identity. This makes PLWH feel helpless. Many people with AIDS, especially in the latter stages of the disease, are weak and become increasingly dependent on others to look after them, which is another source of discomfort. They feel empty and uninterested in things that were previously central to their lives. In terms of physical limitations, people with AIDS, particularly women, may become isolated socially. Some feel guilty for having become infected. In Bangladesh, like in other developing country settings, after diagnosis, patients often fear that they are about to die and worry that no one will come forward to bury them. They also think about their children's future and consider their dream to be broken. Those surrounding them may also be fearful and perhaps suggest that the diagnosis should be kept quiet. The patient and their relatives all need proper counselling and advice for changing their lifestyle.

2.5 Everyday Life and HIV Vulnerability

Individual vulnerability and health risk are part of everyday life among the marginalized people which have been illustrated in this chapter through their 'lifeworlds'. Such biographies provide clear evidence of individuals' social and physical risks which constitute a key role for their potential ill-health, particularly from HIV. Here, poverty, lack of education and employment, poor self-esteem, poor health and nutrition, low social status and exploitation play a role in increasing vulnerability. However, psychological problems also interrupt the individual's daily life. The narratives of commercial sex working women contain examples of their marginalization and the risks that they face every day. The economic, emotional and physical vulnerability of brothel and non-brothel sex workers means that they are easy targets for all manner of abusers and all manner of abuse. They suffer from chronic health problems, which are a result of sexual assault, untreated health problems and overwhelming mental stress. In addition, sex workers' low status in society and uncertainty of income are common issues for them and can push them into risk of HIV infection. The everyday social context of drug users' lifestyles promotes poor health because drug use is not only a serious public health threat, but also a social threat. The physical consequences of the individual's lifestyle mean that they are at risk of a host of diseases, such as hepatitis and HIV. Besides, depression and hopelessness can also affect their individual vulnerability.

Rejection by society and segregation with others after the recovery stage of addiction may lead to the development of subgroup values and customs, which again further reinforce the addictive pattern and the development of deviant, antisocial behaviour. However, with respect to HIV transmission, there is a fear that addicts who are still inhalers might change their mode of addiction to injection when heroin is unavailable. The narratives of PLWH people contain strong traces of the marginalized life that they face every day, including sorrows and anger due to the state policy actions. The 'lifeworlds' of PLWH are a combination of distressing emotions of anxiety, depression and helplessness. Finally, it can be said that poor economic conditions and marginalized social status leave their health vulnerable because of their high identity crisis, stigma and prejudice. As they are less likely to seek appropriate and timely health services, serious health problems like HIV may affect them. HIV is not only a health problem but also a social, economic and cultural issue. In other words, marginalized people's lives are not only a combination of economic and physical vulnerability, but also an effect of social issues like stigma and identity. Their identity and stigmatized status may encourage discrimination and so create a difficult situation. These issues will be discussed in the next chapter.

References

Ahsan, R. M., Ahmad, N., Eusuf, A. Z., & Roy, J. (1999). Prostitutes and their environment in Narayanganj, Bangladesh. *Asia Pacific Viewpoint, 40*(1), 33–44.
Ara, Z. (2005). Violation & denial of access to health-rights for women involved in commercial sex work in Bangladesh. *Women's Health and Urban life, 4*(1), 6–26.
Asthana, S. (1998). The relevance of place in HIV transmission and prevention: The commercial sex industry in Madras. In R. A. Kearns & W. M. Gesler (Eds.), *Putting health into place: Landscape, identity, and wellbeing* (pp. 168–190). Syracuse, NY: Syracuse University Press.
Atherton, J. (2003). *Marginalization*. London, UK: SCM Press.
Barclay, T. (2020). Life takes place: Phenomenology, lifeworlds and place making. *The AAG Review of Books, 8*(1), 3–5.
Barton, M. (1999). Scripture as empowerment for liberation and justice: The experience of Christian and Muslim women in Bangladesh. CCSRG Monograph series 1. Bristol, UK, Department of Theology and Religious Studies, University of Bristol
Bennett, A. (2005). *Culture and everyday life*. London, UK: Sage.
Bloem, M., Hoque, E., Khanam, L., Mahbub, T. S., Salehin, M., & Begum, S. (1999). HIV/AIDS and female street-based sex workers in Dhaka city: What about their clients? In J. C. Caldwell, P. Caldwell, J. Anarfi, et al. (Eds.), *Resistance to behavioural change to reduce HIV/AIDS infection in predominantly heterosexual epidemics in third world countries* (pp. 197–210). Canberra, Australia: Health Transition Center, Australian National University.
Brown, M. (1995). Ironies of distance: An ongoing critique of the geographies of AIDS. *Environment and Planning D: Society and Space, 13*(2), 159–183.
Brown, M. (1997). *Replacing citizenship: AIDS activism and radical democracy*. London, UK: Guilford Press.
Browne, A. W., & Barrett, H. R. (2001). Moral boundaries: The geography of health education in the context of the HIV/AIDS pandemic in Southern Africa. *Geography, 86*(1), 23–36.
Bungay, V., Halpin, M., Atchison, C., & Johnston, C. (2011). Structure and agency: Reflections from an exploratory study of Vancouver indoor sex workers. *Culture, Health & Sexuality, 13*(1), 15–29.
Bungay, V., Johnson, J. L., Varcoe, C., & Boyd, S. (2010). Women's health and use of crack cocaine in context: Structural and 'everyday' violence. *International Journal of Drug Policy, 21*(4), 321–329.
Campbell, J. (1995). HIV and suicide: Is there a relationship? *AIDS Care, 7*(Suppl. 2), S107–S108.
Chaney, D. (2002). *Cultural change and everyday life*. New York, NY: Palgrave.
Chant, S., & Mcllwaine, C. (1995). *Women of a lesser cost: Female labour, foreign exchange, and Philippine development*. London, UK: Pluto Press.
Collins, A. B., Parashar, S., Closson, K., Turje, R. B., Strike, C., & McNeil, R. (2016). Navigating identity, territorial stigma, and HIV care services in Vancouver, Canada: A qualitative study. *Health and Place, 40*(July), 169–177.
Craddock, S. (2000). Disease, social identity, and risk: Rethinking the geography of AIDS. *Transaction of the Institute of British Geographers, 25*(2), 153–168.
Davidson, J. O'. C. (2006). *Prostitution, power and freedom*. Cambridge, UK: Polity Press.
Del Casino Jr., V. J. (2001). Healthier geographies: Mediating the gaps between the needs of people living with HIV/AIDS and health care in Chiang Mai, Thailand. *The Professional Geographer, 53*(3), 407–421.
Dorn, M., & Laws, G. (1994). Social theory, body politics, and medical geography: Extending Kearns's invitation. *The Professional Geographer, 46*(1), 106–110.
Doyal, L. (1995). *What makes women sick: Gender and the political economy of health*. London, UK: MacMillan.
Elmore, K. (2006). The migratory experiences of people with HIV/AIDS (PWHA) in Wilmington, North Carolina. *Health and Place, 12*(4), 570–579.

Eyles, J. (1989). The geography of everyday life. In D. Gregory & R. Walford (Eds.), *Horizons in human geography*. London, UK: MacMillan.
Farley, M. (2001). Prostitution: The business of sexual exploitation. In J. Worell (Ed.), *Encyclopedia of women and gender* (Vol. 2). New York, UK: Academic Press.
Fassi, M. N. (2011). Dealing with the margins of law: Adult sex workers' resistance in everyday life. *Oñati Socio-Legal Series, 1*(1), 36.
Featherstone, M. (1995). *Undoing culture: Globalization, postmodernism and identity*. London, UK: Sage.
Gagnon, A. J., Merry, L., Bocking, J., Rosenberg, E., & Oxman-Martinez, J. (2010). South Asian migrant women and HIV/STIs: Knowledge, attitudes and practices and the role of sexual power. *Health and Place, 16*(1), 10–15.
Gardiner, M. E. (2000). *Critiques of everyday life: An introduction*. London, UK: Routledge.
Gibney, L., Choudhury, P., Khawaja, Z., Sarker, M., & Vermund, S. H. (1999). Behavioural risk factors for HIV/AIDS in a low-HIV prevalence Muslim nation, Bangladesh. *International Journal of STD & AIDS, 10*, 186–194.
Gilbert, L., & Walker, L. (2009). "They (ARVs) are my life, without them I'm nothing"-experiences of patients attending a HIV/AIDS clinic in Johannesburg, South Africa. *Health and Place, 15*(4), 1123–1129.
Goffman, E. (1959). *The presentation of self in everyday life*. New York, UK: Harmondsworth.
Grenfell, P., Platt, L., & Stevenson, L. (2018). Examining and challenging the everyday power relations of sex workers' health. In S. A. FitzGerald & K. McGarry (Eds.), *Realising justice for sex workers: An agenda for change* (pp. 103–122). London, UK: Rowman & Littlefield International.
Guha, M. (2017). Negotiations with everyday power and violence: A study of female sex workers' experiences in Eastern India. Doctoral thesis, University of East Anglia, UK.
Gutierrez-Garza, A (2013). *The everyday moralities of migrant women: Life and labour of Latin American domestic and sex workers in London*. PhD thesis. London, UK: The London School of Economics and Political Science (LSE).
Habermas, J. (1987). *The theory of communicative action, , Vol. 2, Lifeworld and System: A critique of functionalist reason*. Boston, MA: Beacon Press.
Hammond, C. (2008). I'm just here for survival. *The Guardian*, 9th January, London
Harris, P. (2007). *Empathy for the devil: How to help people overcome drugs and alcohol problems*. Dorset, UK: Russel House Publishing.
Hassan, K. M. (2005). Encountering HIV/AIDS. *The Daily Star*, April 20, Dhaka.
Hayes, M. V., Foster, L. T., & Foster, H. D. (Eds.). (1994). *The determinants of population health: A critical assessment*. Victoria, UK: University of Victoria.
Haynes, B. F., Pantaleo, G., & Fanci, A. S. (1996). Toward an understanding of the correlates of protective immunity to HIV infection. *Science, 271*, 24–28.
Heyzer, N. (1986). *Working women in south-east Asia: Development, subordination and emancipation*. London, UK: Open University Press.
Highmore, B. (2002). *Everyday life and cultural theory: An introduction*. London, UK: Routledge.
Holloway, L., & Hubbard, P. (2001). *People and place: The extraordinary geographies of everyday life*. London: Pearson Education.
Hossain, M. M. (2005). *Illicit trafficking in drugs: Perspective Bangladesh, In the publication of Department of Narcotics Control, on the International day against drug abuse and illicit trafficking*. Dhaka, Bangladesh: Government of Bangladesh.
Irwin, S. (1973). A rational approach to drug abuse prevention. *Contemporary Drug Problems, 2*, 3–46.
Islam, S. M. S., Biswas, T., Bhuiyan, F. A., Islam, M. S., Rahman, M. M., & Nessa, H. (2015). Injecting drug users and their health seeking behaviour: A cross-sectional study in Dhaka, Bangladesh. *Journal of Addiction, 2015*, 8 pages.
Jenkins, C., & Rahman, H. (2002). Rapidly changing conditions in the brothels of Bangladesh: Impact on HIV/STD. *AIDS Education and Prevention, 14*(Suppl. A), 97–106.

Kabir, M. A., Goh, K. L., Kamal, S. M. M., Khan, M. M. H., & Kazembe, L. (2013). Tobacco smoking and its association with illicit drug use among young men aged 15–24 years living in urban slums of bangladesh. *PLoS One, 8*, 7. (e68728).

Karner, C. (2007). *Ethnicity and everyday life*. London, UK: Routledge.

Kearns, R. A. (1993). Place and health: Towards a reformed medical geography. *The Professional Geographer, 45*(2), 139–147.

Kearns, R. A. (1994). Putting health and health care into place: An invitation accepted and declined. *The Professional Geographer, 46*(1), 111–115.

Kellehear, A., & Cvetkovski, S. (2004). Grand theories of drug use. In M. Hamilton, T. King, & A. Ritter (Eds.), *Drug use in Australia: Preventing harm* (pp. 53–63). Australia: Oxford University Press.

Kempadoo, K., & Doezema, J. (Eds.). (1998). *Global sex workers: Rights, resistance, and redefinition*. New York, NY: Routledge.

Khan, M. A., & Khanum, P. A. (2000). Influence of son preference on contraceptive use in Bangladesh. *Asia Pacific Population Journal, 15*, 43–56.

Khan, S. (1988). *The fifty percent: Women in development and policy in Bangladesh*. Dhaka, Bangladesh: University Press Limited (UPL).

Khan, Z. R., & Arefeen, H. K. (1989). *Potita nari: A study of prostitution in Bangladesh*. Dhaka, Bangladesh: Dhaka University.

Khanam, R., Reza, M., Ahmed, D., Rahman, M., Alam, M. S., Sultana, S., et al. (2017). Sexually transmitted infections and associated risk factors among street-based and residence-based female sex workers in Dhaka, Bangladesh. *Sexually Transmitted Diseases, 44*(1), 22–29.

Knight, V. C. (2006). *Drug users at risk to HIV: Documenting our experience 2000–2005, HIV program*. Dhaka, Bangladesh: Care Bangladesh.

Koester, S., Rahman, K. Z., Kelly, R., & O'Day, T. (2005). *Drug sharing and injecting networks in Bangladesh: Implications for HIV transmission, In the publication of Department of Narcotics Control, on the International day against drug abuse and illicit trafficking*. Dhaka, Bangladesh: Government of Bangladesh.

Kumar, N., Raju, S., Atkins, P. J., & Townsend, J. G. (1997). Where angels fear to tread? Mapping women and men in India. *Environment and Planning A, 29*(12), 2207–2215.

Lex, B. W. (2000). Gender and cultural influences on substance abuse. In R. M. Eisler & M. Hersen (Eds.), *Handbook of gender, culture and health* (pp. 255–297). Hillsdale, NJ: Lawrence Publishers.

Lorway, R., Lazarus,L., Chevrier, C., Khan, S., Musyoki, H. K., Mathenge, J., Mwangi, P., Macharia, P., Bhattacharjee, P., Isac, S., Kimani, J., Gaaki, G., Becker, M., Moses, S. and Blanchard, J. (2018) Ecologies of security: On the everyday security tactics of female sex workers in Nairobi, Kenya, Global Public Health, 13 (12), pp 1767-1780

Luginaah, I. (2008). Local gin (akpeteshie) and HIV/AIDS in the Upper West Region of Ghana: The need for preventive health policy. *Health and Place, 14*(4), 806–816.

Marques, A. C. (2019). Displaying gender: Transgender people's strategies in everyday life. *Symbolic Interaction, 42*(2), 202–228.

Mau, S. (2010). *Social Transnationalism: Lifeworlds Beyond The Nation-State*. Oxon, UK: Routledge.

May, J. (1996). Globalization and the politics of place: Place and identity in an inner London neighbourhood. *Transactions of the Institute of British Geographers, 21*(1), 194–215.

McNeil, R., Shannon, K., Shaver, L., Kerr, T., & Small, W. (2014). Negotiating place and gendered violence in Canada's largest open drug scene. *International Journal of Drug Policy, 25*(3), 608–615.

Morgan, D. (2004). Everyday life and family practices. In E. B. Silva & T. Bennett (Eds.), *Contemporary culture and everyday life*. Durham, NC: Sociology Press.

NASP. (2003). *HIV in Bangladesh: Is time running out? National AIDS/STD Programme (NASP), Ministry of Health and Family welfare*. Dhaka, Bangladesh: Government of Bangladesh.

O'Brien, C. P., & McLellan, A. T. (1996). Myths about the treatment of addiction. *Lancet, 347*, 237–240.
Patterson, C. B. (2015). Beyond the stigma: The Asian sex worker as First World savior. In M. Laing, K. Pilcher, & N. Smith (Eds.), *Queer Sex Work* (pp. 53–65). Oxon, UK: Routledge.
Pullen, K. (2005). *Actresses and whores: On stage and in society*. Cambridge, UK: Cambridge University press.
Rahman, M., & Zaman, M. S. (2005). Awareness of HIV/AIDS and risky sexual behaviour among male drug users of higher socioeconomic status in Dhaka, Bangladesh. *Journal of Health, Population and Nutrition, 23*(3), 298–301.
Rahman, M., Zaman, M. S., Sekimoto, M., & Fukui, T. (2000). HIV-related risk behaviours among drug users in Bangladesh, letter to the editor. *International Journal of STD & AIDS, 11*, 827–828.
Rawat, A., & Kumar, S. (2015). Lifeworlds of children of sex workers in Budhwar peth. *The Indian Journal of Social Work, 76*(4), 579–596.
Report, U. B. I. N. I. G. (1995). *Trafficking in women & children: The cases of Bangladesh*. Dhaka, Bangaladesh: Unnayan Bikalper Nitinirdharoni Gobeshona (UBINIG).
Revenson, T. (2001). Chronic illness adjustment. In J. Worell (Ed.), *Encyclopedia of women and gender* (Vol. 1). New York, UK: Academic Press.
Rhodes, T. (2009). Risk environments and drug harms: A social science for harm reduction approach. *International Journal of Drug Policy, 20*(3), 193–201.
Seamon, D. (2015). *A Geography of the lifeworld: Movement, rest and encounter*. New York, NY: Routledge, Taylor and Francis.
Seib, C., Fischer, J., & Najman, J. M. (2009). The health of female sex workers from three industry sectors in Queensland, Australia. *Social Science and Medicine, 68*(3), 473–478.
Sharma, M., Rahman, K. Z., Kelly, R. and O'Day, T. (2006). *Reducing the demand for drugs and preventing HIV in Bangladesh: A partnership between law enforcement, the community and treatment agencies, In the publication of Department of Narcotics Control, on the International day against drug abuse and illicit trafficking*. Dhaka, Bangladesh: Government of Bangladesh.
Shirin, T., Ahmed, T., Iqbal, A., Islam, M., & Islam, M. N. (2000). Prevalence and risk factors of hepatitis B virus, hepatitis C virus, and human immunodeficiency virus infections among drug addicts in Bangladesh. *Journal of Health, Population and Nutrition, 18*(3), 145–150.
Singh, N. S. C., Sivek, C., Wagener, M., Hong Nguyen, M., & Yu, V. L. (1996). Determinants of compliance with antiretroviral therapy in patients with human immunodeficiency virus: Prospective assessment with implications for enhancing compliance. *AIDS Care, 8*(3), 261–269.
South, N. (Ed.). (1999). *Drugs: Cultures, controls and everyday life*. London, UK: Sage.
Stone, L. D. (2013). *Self-Emplacement in the lifeworld: The geographic imagination of American middle adolescents*. PhD dissertation, Texas State University.
Sutherland, I. (2004). *Adolescent substance misuse: Why one person may be more at risk than another, and what you can do to help*. Dorset, UK: Russell House Publishing.
Tahmina, Q.-A., & Moral, S. (2004). *Sex-workers in Bangladesh-livelihood: At what price?* Dhaka, Bangladesh: SHED.
Takahashi, L. M., & Smutny, G. (2001). Explaining access to human services: The influence of descriptive and behavioural variables. *The Professional Geographer, 53*(1), 12–31.
Taposh, S. (2006). Continued exploitation deters rehabilitation of sex workers. *The Daily Star*, Issue 256, September 30, Dhaka.
Thompson, C. (1996) The genes that keep AIDS at bay. *New Scientist*, *16*, 6th April.
Travis, C. (2017). GeoHumanities, GIScience and smart city lifeworld approaches to geography and the new human condition. *Global and Planetary Change, 156*(September), 147–154.
Ullah, A. K. M. A. (2005). Prostitution in Bangladesh: An empirical profile of sex workers. *Journal of International Women's Studies, 7*(2), 111–122.
UNAIDS. (2002). *Gender and AIDS*. Geneva, Switzerland: Joint United Nations Programme on HIV/AIDS (UNAIDS).

UNDP. (2012). *Sex work and the law in Asia and the Pacific: Laws, HIV and human rights in the context of sex work*. New York, NY: United Nations Development Program.
van Blerk, L. (2016). Livelihoods as relational Im/mobilities: Exploring the everyday practices of young female sex workers in Ethiopia. *Annals of the American Association of Geographers, 106*(2), 413–421.
Walklate, S. (Ed.). (2007). *Handbook of victims and victimology*. Portland, OR: Willan Publishing.
Watts, M. J. (1991). Mapping meaning, denoting difference, imagining identity: Dialectic images and postmodern geographies. *Geografiska Annaler, 73*(1), 7–16.
Whelan, G. (2004). The pharmacological dimension of psychoactive drugs. In M. Hamilton, T. King, & A. Ritter (Eds.), *Drug use in Australia: Preventing harm* (pp. 17–32). Australia: Oxford University Press.
Wilkie, L. A. (2001). Race, identity and Habermas's lifeworld. In C. E. Orser Jr. (Ed.), *Race and the archaeology of identity*. Salt Lake City, UT: The University of Utah Press.
Williams, M., Bowen, A., Ross, M., Freeman, R. C., & Elwood, W. (2000). Perceived compliance with AZT dosing among a sample of African-American drug users. *International Journal of STD & AIDS, 11*(1), 57–63.
Wilton, R. D. (1996). Diminished worlds? The geography of everyday life with HIV/AIDS. *Health & Place, 2*(2), 69–83.
Wilton, R. D. (1999). Qualitative health research: Negotiating life with HIV/AIDS. *The Professional Geographer, 51*(2), 254–264.
Young, L., & Ansell, N. (2003). Fluid households, complex families: The impacts of children's migration as a response to HIV/AIDS in Southern Africa. *The Professional Geographer, 55*(4), 464–476.

Chapter 3
Stigmatized People and Societal Prejudice

3.1 Introduction

Stigma is not only an isolated sociological concept. It is closely wrapped up with many other aspects of the human condition, leading to discrimination and marginalization. The term 'stigma' refers to an attribute that serves to 'discredit a person or persons in the eyes of others' (Franzoi, 1996; p. 403) or 'devalues the person' (Hopper, 1981); it can be seen as a 'principled refusal' (Fischhoff, 2001) or in terms of 'socially disqualifying' attributes (Katz, 1981). Kasperson, Golding, and Kasperson (2005) define stigma 'as a mark placed on a person, place, technology, or product, associated with a particular attribute that identifies it as different and deviant, flawed or undesirable' (p. 171). Parker and Aggleton (2003) argued that stigma is a social process that produces and reproduces inequalities and, in this case, stereotyping may be one mechanism through which the process takes place. Stigma in Erving Goffman's (1963) terms is an undesirable differentness from what the non-stigmatized have anticipated. According to Goffman, there are three different types of stigma: first, physical disfigurement, deformities and disease; second, aberrations of character and personality and third, social categorizations such as race, nationality and religion. A stigmatized act is unacceptable whatever the associated benefits. An important aspect of stigma is that stigmatized people often themselves accept the concept of stigma that is current in their own culture. In other words, stigma is very often also 'self-stigma', an attribution which the despised minority internalizes (Goffman, 1963).

In Goffman's conception of stigma, there is the implication that the attributions spoil the person's identity permanently. The experience of stigma has a profound effect both in its emotional impact for the individual concerned and in its social repercussions for the marginalized group as a whole. At an individual level, the impact of stigma and social exclusion can be devastating, leading to low or diminished self-esteem (Hogg, 1985), poor social relationships, isolation, depression,

A. Paul, *HIV/AIDS in Bangladesh*, Global Perspectives on Health Geography,
https://doi.org/10.1007/978-3-030-57650-9_3

self-harm (Mason, Carlisle, Watkins, & Whitehead, 2001) and feelings of loss of control, embarrassment and deficiency (Benjamin, 2001). There is a close interplay of psycho-social factors contributing to the changing dynamic of stigmatized groups and individuals in society (Franzoi, 1996). Stigmatized groups are, by definition, devalued in and by society (Hogg & Vaughan, 2002), have relatively low status and little power, and find it difficult to avoid society's consensual negative image of them. In other words, victims are frequently marginalized, sometimes completely ignored and further victimized as a result of the responses to their victimization (Williams, 1999). In general, although some stigmatized individuals are vulnerable to low self-esteem, diminished life satisfaction and in some cases depression, most members of stigmatized groups are able to weather the assaults and maintain a positive self-image (Crocker, Major, & Steele, 1998).

Moreover, 'Identity' is a commonly used term in a number of different ways which can shake the foundations of our lives in everyday discourse (Jenkins, 1999). It is a relational construction and relative to situations and contexts which have both temporal and spatial dimensions (Holzner & Robertson, 1980). Identity only becomes an issue when it is in crisis, when something assumed to be fixed, coherent and stable is displaced by the experience of doubt and uncertainty (Mercer, 1998). An individual's identity is formed and maintained in the course of interaction with others, because we all need the 'positive regard' of others and strive to obtain from others confirmation of our view of ourselves (Blacking, 1983). Erikson observed that every person's 'psycho-social identity' contains both positive and negative elements (Erikson, 1968). These negative elements or 'identity crises' caused or reinforced by stereotypes held by members of other groups can lead to self-hate. In an 'identity crisis', a person (e.g. sex worker or drug user) is denied easy access to her/his relatives or neighbours due to involvement in 'unsocial' work or with unsavoury elements of society (Butler, 2004). This can lead to a deep sense of displacement, a sense of not knowing or not belonging to the 'social world' (Lawler, 2008). However, discrimination is the behavioural expression of prejudice (Hogg & Vaughan, 2002). A person's experience of discrimination is most acutely felt at the individual level in their day-to-day interactions with others (Bennett, 2005). HIV/AIDS is an illness associated with perceptions of stigmatized sexual acts and illicit drug use. Sontag (1989) pointed out that the stigma of AIDS elicits fear and avoidance rather than sympathy. Ghosh, Wadhwa, and Kalipeni (2009) found that various socio-economic and cultural aspects create an environment of vulnerability to HIV/AIDS.

In many societies, AIDS is seen as a disease of shame and a number of HIV/AIDS infected people have faced discrimination in terms of medical care, and have been rejected by family and friends or forced to leave their occupation (Muyinda, Seeley, Pickering, & Barton, 1997). HIV infection and AIDS draw a particularly negative societal response, which can result in individual suffering, and group marginalization (Clarke, 2001). As yet there is no cure for AIDS, and this uncertainty remains a strong influence on the way in which HIV and AIDS impact psychologically on those people who are infected and affected (Carlisle, 2001). The lack of a cure also affects the way in which the virus is viewed by society (Carlisle, 2001). Despite the biological vulnerabilities, the AIDS epidemic also exposes hidden social vulnerabilities in the human condition, such as stigma and discrimination. Discrimination is one such reaction which faces people affected by HIV, and stigmatizing societal responses are the product of complex beliefs, many of which are rooted in views around sexuality and sexually shared infections (Bennett, 2005). Ulasi et al. (2009) identified four HIV/AIDS-related stigma and discrimination especially employment, screening, disclosure and social contact. Gilbert and Walker (2009) illustrated how fear of stigma play role in patients experiences. Craddock (2012) discussed tuberculosis which poses one of the biggest threats to individuals living with HIV in most low-income regions of the world. Ransome, Kawachi, Braunstein, and Nash (2016) found that structural inequalities drive late HIV diagnosis. Scambler (2020) considered stigma as way of vulnerability. He discussed elaborately how money, power and media convert shame into blame in the process of stigmatization.

In order to study the objective of marginalized people's stigmatized lives, factors such as stigma, identity, rights and suffering caused by discrimination have been explored in this chapter. These marginalized people (commercial sex workers; drug users, mainly heroin users; and people living with HIV) are struggling every day for their economic, social and physical survival and lead a very stigmatized life, without access to health care services, little knowledge about their rights and a feeling of hopelessness. In this research, the identity of the sex workers, drug users and PLWH is illustrated, particularly how they are treated by their families, villagers, neighbours or health workers and how their spoiled identity is affecting their lives.

3.2 Sex Workers' 'Bad' Identity

Sex work is itself a socially stigmatized matter (Patterson, 2015) and having various stigmas in sex work and brothel settings (Blithe, Wolfe, & Mohr, 2019). The term 'sex worker identity' is commonly used to refer to the imagined or portrayed personage in the women's stories. Many of them consider their own profession as 'bad', demeaning or shameful (Chant & Mcllwaine, 1995). A woman who has engaged in sexual activity outside the protective family framework is in a very vulnerable position and she may be rejected by her family (Goddard, 1993). Women working as prostitutes are perceived as bad girls, suffering stigma and increasingly criminalized by the state (O'Neill, 1997). They may also choose to describe themselves in a variety of ways: as escorts, masseuses, working women or simply as prostitutes (Wilton, 1994). Women who continue to live in regular households but who sell sex without the knowledge of their families and neighbours maintain a close secrecy about their working lives (Asthana, 1996). A prostitute's subsequent life is spent almost entirely with her fellow workers and clients, for her parents dare not keep in touch because of fear of social ostracism (Blanchet, 1996). Phoenix (2000) constructed a few contingent elements of the 'prostitute identity', for example, prostitutes as 'workers', 'commodified bodies', 'loving-partners' and 'victims'. Bungay, Halpin, Atchison, and Johnston (2011) discussed sex workers lives and their exploitation and contradictory attitude on own work. Fassi (2011) discussed the sex workers' understandings and associated practices of resistance. Gutierrez-Garza (2013) found their everyday challenges due to their status and experiences of illegality. Patterson (2015) depicted the sex worker as an important and essential function of development and globalization. Rawat and Kumar (2015) discussed their perception on their 'self' and aspirations. Walker (2017) detailed their wide-spread stigmatization especially in accessing basic services, for example, healthcare and childcare.

Today, some or all aspects of prostitution are illegal in all countries (O'Neill, 1997) and most societies define the prostitute as a 'criminal' and many people care little about either the violence that is committed against prostitutes or treating them as ordinary human beings (Alexander, 1996). Fassi (2011) mentioned sex worker's activity which is not considered as legal but not illegal either. Patterson (2015) emphasized on necessity of de-stigmatize sex work as a legitimate form of labour. Lorraine van Blerk (2016) also discussed the necessity for work security, safety and service access for sex workers. To date there is no legislation that recognizes commercial sex as a profession. Instead sex workers are considered to be fallen women who do not deserve social dignity or recognition (Sultana, 2015). As a marginalized group with no legal recognition and inadequate protection, the rights of commercial sex workers in Bangladesh are often violated (Ara, 2005; Ullah, 2011). Every person involved in the profession has become a victim of social exclusion and marginalization (Chowdhury, 2006; Paul, 2009). Even death fails to end the sex workers' misery. Regardless of whether a sex worker is Hindu, Muslim or Christian, she is still denied basic funeral rites. However, due to social taboos and widespread prejudice, prostitution has not yet been identified as an economic problem of a serious nature (Khan, 1988). Here the identity of sex workers is analysed in terms of the ways in which they are publicly discussed, socially treated and officially depicted.

Sex Workers 'Identity Crises' Commercial sex workers in Bangladesh generally try to hide their profession from their family, neighbours and village community. They face an identity crisis and introduce themselves as a factory or garment worker, nurse, family planning worker or NGO worker to protect their status and maintain a good image. Almost all sex workers have a tendency to change their names when they enter the profession in order to conceal their 'real' identity. But some sex workers believe that their mother may have guessed their profession. One of the hotel sex workers I interviewed in Jessore, Nipa said that '*my mother may know because she can read her child's mind*'. But a casual girl, Lota, thinks that although her mother may assume she is a sex worker from her earnings, she has never asked about it due to the possible disgrace, but she sometimes asks 'why I didn't come to home last night'. On the other hand, one of the Baniashanta sex workers, Salma told me that her mother knows her profession. As she is the only earner in their family, her mother is compelled to accept this. Her mother sometimes comes to the village adjacent to the brothel and collects money from her daughter. Salma also visits their house in Khulna city and stays with her mother for a few days. Her mother emotionally asks her not to return to the brothel again. She consoles her mother because she needs to earn money, otherwise neighbours or relatives will spread 'stories'. She believes that money can remove the disgrace: '*When people see money then they will not say anything, even if I do some bad things in our house nobody will ask anything*'. Sex workers, particularly brothel girls, may introduce their lover or *babu* as their husband to their relatives and neighbours in order to overcome the marital status problem, which is an issue for women in Bangladesh. Alternatively, some sex workers like Lota are less concerned about their stigmatized identity. When neighbours make any odd comments, she just ignores them. Although she feels that she cannot answer back, she does not care about their remarks.

Sex workers, particularly brothel girls, believe that society has a double standard towards them. On the one hand, male relatives who know about their profession, particularly brothers or uncles, threaten them with violence or death if they stay at the brothel. On the other hand, these relatives take money, although they refuse acknowledgement for fear of losing status. Many family members of sex workers receive gifts from them, but villagers or neighbours do not know the source. Also, according to the girls, the biggest hypocrisy is that the kind of men who commonly criticize them are also their customers. I met with a girl named Bristi who is working in the Jalaipotti brothel of Jessore. She told me about her family problems:

> 'In our village, no relatives recognize me. They have forbidden me to go home because I am living in the brothel. Recently one of my family members died but none of my family informed me and I got the information two weeks later. When I asked why, I was told that if I had attended the villagers would not have buried your relative because of your profession. I felt very sad but I realised that my presence would create embarrassment for my parents. Already the villagers do not mix with them properly because of me. But if I leave the brothel forever for the sake of family status, I would lose on both sides. My family would not accept me and it would be difficult to return here again'.

Sometimes casual sex workers like Parvin, Fatema and Johura entertain customers in their own houses by introducing them as relatives or colleagues but they face many problems with neighbours, particularly women, who make comments. They need to give many explanations to maintain their reputation as a 'good girl'. All the time sex workers need to be conscious about concealing their 'bad' identity. Sometimes, local *mastaans* threaten the girls, telling them to leave, or they demand money. However, casual sex workers who have (ir)regular husbands always need to hide their profession. One of the hotel-based girls, Keya told me that her husband knows that she is employed by an NGO that works with hotel girls. So, she has no fear of going to the hotel. But when she has sex with her husband she practises some techniques 'so that he won't guess about her outside sex'. Another sex worker told me that her husband is sometimes suspicious about her character. Then a 'soft reminder' is given by the wife, as her husband depends on her income.

Adjacent villagers' perceptions or attitudes towards Baniashanta brothel girls are mixed. Some consider them to be '*kharap meye*' (lost girls) but many like the brothel girls. Moushumi told me that they mix freely with the village people. In the past, school teachers were rude to their children and villagers obstructed them from collecting water but now everything has been changed because of NGO pressure and advocacy meetings with different groups. Also, one of the old members of this brothel, Nargis thinks that villagers are benefiting economically from the brothel girls because they can sell their vegetables and fish and the older women can work in the sex workers' houses as a maids or cooks. Villagers can sell water, and if any villagers face hardship, the brothel girls help them. Efforts to reduce the stigma attached to sex work might eventually allow sex workers to maintain ties with their families and hope for a better future.

Self-Estimation About 'Sex Work' Although society's tolerance or acceptance of prostitution has varied at different times and places, there is no place and no time that has it has been entirely without stigma or repression. A woman who has engaged in sexual activity outside the family is in a very vulnerable position in the third world countries and she may be rejected by her all family members. In Bangladesh, most women and girls are being driven to this profession through trafficking. There is no legislation that recognizes commercial sex as an activity and sex workers are considered as fallen women who do not deserve social dignity or recognition. They are subjected to exploitation and violence, and their activities beyond the law limit their access to the very health and other services which might serve their health and safety needs and the health of their clients. As a result, many negative feelings have developed among sex workers. Most have low self-esteem and feel that they cannot go out freely and mix with other people on the outside because of stigma. As a result, inferiority complexes develop in their minds and ultimately they start to think that they are alone in the world. So, they want the recognition of sex work as 'work'. Their logic is that they did not come to the brothel willingly and that this work was forced on them. One of the sex workers in the Jessore Maruari Mandir brothel, Bobita told me that in this respect '*the state should give us the official status of workers*'.

With regard to self-esteem, brothel girls have a variety of views. Some in an FGD at Baniashanta said that they are doing bad work and that they feel they are 'lost girls'. They consider 'good work' to be employment in a garment workshop, a beauty parlour, embroidery or handicrafts. But one countered this: '*is there anything written down on our bodies to say that we are bad women? If we go to another place, how will people know us as bad?*' Moushumi was even more positive: '*some people want to say that sex work is a bad thing, but I don't consider it as such because you will not give me 10 taka if I do not give you pleasure. We do not steal money from anyone; we earn through labour*'.

Regarding the brothel, many girls consider it as a 'hell' or 'prison' but they have to stay there in order to buy food to live. Baby told me that '*I feel always sad, have many sorrows, I don't talk with others. Here the girls joke with each other but I cannot take part. Basically I don't like this environment but what could I do on the outside? I would still need to do sex work, so it would be the same*!' On the other hand, some girls consider the brothel to be a good place to work independently, without disturbance, with freedom. One girl, Kazol told me at the Jessore brothel that '*I consider this place as a peaceful place, for food, for clothes, recreation. I have a TV, a CD, everything and importantly we are safer here than outside*'.

However, many girls also explained that they have no interest in working in a garment factory because, firstly, it is an under-waged and labour-intensive job; and secondly, many girls are forced into sex with their immediate boss without pay or respect.

In terms of sex workers' incomes, there are some beliefs among the girls about honest and dishonest income. Many believe that they are earning the money in a dishonest way, so their money will somehow dwindle away within a short time. On the other hand, if they could earn money in an honest way, for example, doing a 'respectable' job, then the money would bring blessings or abundance to them. Many sex workers consider their work as a 'sin' and their logic is that it is forbidden in the holy book. They are confused whether Allah will forgive them for this kind of '*kharap kaz*' (bad work). One home-based sex worker, Dibba told me that '*I am trying to overcome this sin as it's a dishonest income. I cannot say to anybody about my earnings. Allah has forbidden this type of work and income, so I am saying it is dishonest money*'. I found one casual sex worker, Priyanka, who believes that, due to her sin, her baby daughter has been suffering from sickness and there is no improvement, but now the money she earns is going fast. '*I always think that I am doing sinful work, I am feeding my baby with sinful money, I feel very bad with myself; she is always sick*'.

Sex workers continue to suffer stigma and abuse particularly at the hands of those who should help, such as police officers. The constitution of the People's Republic of Bangladesh says that 'the state shall adopt effective measures to prevent prostitution and gambling' (Article 18:2). However, there is also provision for an adult woman to take up prostitution by making an affidavit in a first class magistrate's court with a notary public. This provision is proof enough that the state does not prohibit the sale of sex. To some extent, these women sense entrapment, as they feel 'forced' to choose a profession that is highly stigmatized and not regarded positively by society due to poverty and lack of educational background. This can result in high levels of stress.

Rights and Empowerment In Bangladesh, most of the women who are bound to sex work have little or no rights. Sex workers think that their entrance in this 'profession' itself is the result of continuous violation of human rights in the country. They are being harassed and tortured by law-enforcing agencies due to a lack of transparency legally about their profession. The social status of sex workers in Bangladesh is considered so low that in the past they were not allowed to wear shoes/sandals when leaving the brothel, nor even the typical dress (*shalwar-kameez*). Preventing sex workers from wearing sandals and *shalwar-kameez* enables society easily to identify and ostracize them.

To counter the historic stigmatization, some NGOs in Bangladesh have been implementing various programmes for sex workers. NGO officials believe that in the last 10 years much of the stigma associated with prostitution has been broken down by the increasingly positive attitude of government and the public. In addition, in some brothels, sex workers are organizing themselves into self-help groups for mutual welfare and to protect themselves from harassment. Some NGOs have started organizing and encouraging sex workers to develop self-help groups which are working as alliances with other similar organizations. Despite such positive initiatives, most sex workers still find it difficult to access public services, such as education and healthcare. Some brothel girls told me that in the past they did not understand anything about their rights. Now their eyes have been opened and they are more united. They also feel that now they can answer back when government hospital doctors refuse to provide treatment in a timely manner, or when school teachers want to know about the father's identity for their children at school admission. They now understand that they have the right to enrol their children and that the mother's identity is enough. Rotna told me:

> 'I didn't come in the brothel from my mother's womb. Someone forced me to join. Actually previous girls were foolish and scared about their identity and rights, but now if anybody challenges us, we can argue because we have been trained by the NGOs, and now we understand that we have the right to go to a bank and open an account, the right to buy land and build a house, the right to go to a big function, or to go to a 'milad mahfil' (religious meeting) because Allah also created us'.

Due to different rights-based approaches and empowerment efforts by different NGOs, sex workers are now able to raise voices on different issues. Although there are still *mastaan* and police problems, many girls believe that 'the law is equal for all'. Regarding burial rights, I heard that in the past when someone died at Baniashanta brothel, they were not able to organize a religious ceremony, and there was no system for burying or '*janaza*' or prayers. Recently, they have been given a piece of land by the government for a graveyard due to pressure by NGOs. Moushumi told me '*when infants died at birth in Baniashanta brothel we could not arrange any religious ceremony for the dead bodies, we had to bury them on a "char" or barren island or float them on the river current. When we went to any "hujur" for religious involvement, they refused, but that baby was innocent. We may guilty for bringing the baby into this world, but they never heard our requests, and always blamed us in the name of religion*'. In this connection, in Khulna I asked an Islamic cleric, Shajahan (not his real name) that in Islam, there is a '*fotowa*' or Islamic law that when a girl is involved with commercial sex openly then she loses the '*iman*' or faithfulness. As a result, they are not allowed burial according to Islamic law. But he thinks that as a human being they can get burial rights.

3.3 Stigmatized Drug Addiction

Goffman (1963) suggests that stigmatization is a complex process of social interaction between stigmatized and non-stigmatized persons with certain culturally salient characteristics, which leads to social consequences of reduced opportunities, discrimination and even outright rejection. 'When we come across terms such as "drugs" or "addiction", we all bring to it a limited 'common-sense' view of these concepts that are shaped by our culture, the media, our own prejudice and other factors' (McDermott, 1992; p. 195). The social meaning of drug use differs across time and geographical space. The relationship between drug abuse and social deprivation holds in developed as well as developing countries (Currie, 1993). Drug users' lifestyles may additionally be harmful because of the associations that are bound to be made to obtain an illegal drug such as heroin (Robertson, 1987). The social and economic marginalization of drug user populations which combines to produce a shared sense of social suffering may in turn reinforce close social bonds within networks which 'act as the conveyor belts of drug injection technical knowledge and encouragement' (Singer, 2001; p. 205). The general poor health, and social and material conditions of many drug users complicate the relationship between substance use and ill-health. Consequences of substance abuse occur throughout the life cycle and include compromised health, family malfunctioning and poor child welfare, as well as increased crime and incarceration rates (Lex, 2000).

The negative effects of the substance may involve impairment of physiological, psychological, social or occupational functioning (Lewis, Dana, & Blevins, 2002). However, it is clear that the drug user is living in a violent or potentially violent world. The life is one of risk and confrontation with peers, criminals, police and society and therefore violent acts are commonplace. Most cultures have adopted a largely punitive approach to heroin use. Life as a drug addict is a breeding ground for other assaults such as violence, poverty, homelessness, poor familial relations or abandonment (Zierler, 1997). Problem drug use can seriously affect families, leaving those involved floundering in a sea of anger, frustration, fear and isolation (Barnard, 2007). In terms of the link between specific illicit substances and criminality, the association with the longest historical pedigree is the link between opiates and crime (Carnwath & Smith, 2002). The development of drug problems in a close family member was often an insidious process marked by small, but significant, changes in manner, behaviour and appearance (Usher, Jackson, & O'Brien, 2005). Once the family became aware that there was a problem with drugs, the most likely reaction was utter panic, arising from a lack of knowledge and experience.

Heroin has been referred to as the 'devil drug', synonymous with addiction and crime (Miller, 1994). The major consequence of this criminalization/stigmatization is causing addicts to conceal the fact that they use drugs (McDermott, 1992). The idea that heroin use leads to crime is put forward today almost as a matter of common sense. The popular argument is that people turn to property crime (and may employ violence) in order to support their heroin habits (Dorn & South, 1987). In qualitative work on drug users, Ware, Wyatt, and Tugenberg (2005) found that stigmatization and stereotyping may contribute to unequal treatment for drug users and other populations who are living with HIV. Therefore, we have a large number of people, engaging in a very dangerous form of behaviour, and with a high potential for inflicting unnecessary damage on themselves and others. Illicit drug use is highly stigmatized and virtually all habitual drug users in Bangladesh are marginalized and discriminated against (Khosla, 2009; Knight, 2006). In Bangladesh, the term 'drug user' is used to refer to those who ingest illegal psychoactive substances. But it is also a stigmatized term for those who are unable to sustain their habit financially, fulfil commitments to others and to society at large and retain control over life circumstances. However, the stigma of drug use is compounded for users who inject drugs and have risky behaviours. In describing drug users' stigmatized worlds, my research stresses their living place, their inadequate home life, their continued use of drugs as part of ordinary social relationships and their generally hopeless and futile future (some photos of the drug users are shown in Photo Series 3.1 as symbolic). In the following section, addicts' stigmatized lives, their identity and related prejudice that surrounds them will be described through a number of case studies.

Addicts' Blamed Lives A stigmatized person's social values may decrease to levels below what one would expect taking into account the risks associated with it. In Bangladesh, there are views about good and bad drugs. People no longer mind about cannabis, although it was considered a 'bad addiction' a few years ago. Recently, young people have started taking stimulating syrup of Indian origin, *phensidyl*, containing codeine phosphate and ephedrine, which was originally used as a medicine for coughs and colds but due to its misuse it was banned a few years ago. In fact, nearly 80% of the drugs used by IDUs are easily available at chemist shops in Bangladesh and in the South Asian region at large.

Many drug users have lost touch with their own families and with mainstream society due to their involvement in different crimes including burglary and blackmail. When their *bera* or withdrawal symptoms appear, they become desperate and some addicts steal goods from their own house or make demands or threats to their parents and other family members to get drug money and for this reason many are

Photo Series 3.1 Drug user's life with sorrows

abandoned by their relations. When they start using heroin, they become a disruptive element in the family and are introduced to the neighbours as dishonoured. The addiction locks them into a cycle of transgression which makes it difficult to get proper guidance and to break out of a lifestyle which makes them vulnerable. I talked with wives of recovered drug addicts who suffered much from their husbands during their addiction. One of them, Sabina, said that:

> 'My husband never tried to understand me. He always looked for my savings and he even sold our son's wrist watch for heroin money. When his addiction took hold he had no sense of what is right and wrong. Many times he borrowed money in my name and I had to pay these debts. Sometimes the lenders would beat him for their money and I had to rescue him. He even sold our mosquito net. He hit me when I tried to stop him taking heroin. He never thought of spending money on us, and only thought of heroin'.

Drug addicts are stigmatized their whole lives. Sometimes they spend time in restaurants or in other places to experience the drug's effects as they have no shelter or a good place to take a proper sleep. They are asked to leave when they are recognized and they are not able to protest. They always face blame and many drug addicts want to commit suicide. In Bangladeshi, society addicts have to face insults and physical abuse from people in general and they lack close friends because nobody trusts them. If anything negative happens in an area, people always blame addicts. Addicts are victims of discrimination. I found one in Jessore railway station, named Shapon. He tearfully told me about his addicted life:

> 'I don't like this life, because nobody likes us. People insult us wherever we go. For example, if any dog comes to you, you will feed to the dog, but if I come along, you will not give me any food, and you may even beat me. So, a dog is better off than a heroin addict. You see, my clothes are dirty, I cannot sleep and take food properly, but sometimes I want to practise 'namaz' (prayers) for purifying myself, but if I go to the mosque, people will consider me a thief due to my identity as heroin addict'.

Drug addicts lose respect in society. The stigma may affect other members of the family. For instance, an addicted parent may find that no family will come forward for his or her daughter to marry. I found one female addict in Jessore, Jomila who has two sons. But when she became addicted, local people forced her two sons to leave their mother. The concern was that they would also become addicts. Drugs make the addict hopeless in the eyes of society and their independent voice disappears. The hatred of heroin addicts is so universal that even small children blame them. During my interviewing of drug users at Jessore rail station, local people warned me that they would steal from me. When I was talking with one addict, a woman came close to us and thought that I was from a government department and would punish the heroin smokers. She shouted that they should be jailed or killed because they are a liability to society and the country. They steal state property like steel plates from the railway line to sell to raise the money for heroin.

Addicts' Negative Identity and Impact Drug addicts have such a negative identity that most have no access to, or acceptance by, their parents and other family members. I found one female addict at Jessore, Rabeya, who has been facing denial by her brothers. As she takes drugs, she cannot afford to wear good clothes, and roams around the town collecting rubbish to manage the drug money, so her brothers do not recognize her as their sister. She is not allowed take even a glass of water from their house. In most cases, even close family members do not trust them. In one participant observation in an NGO-run drug treatment centre in Jessore, I found one drug patient, Asadul who told me a story of mistrust by his own younger brother and friends:

> 'My younger brother asked me to buy some academic books for him. I went to one of my friends and requested his help to purchase these books. But my friend said that you are involved in heroin addiction, so I don't believe you. Bring your younger brother with you. When I mentioned this to my younger brother, he felt embarrassed and puzzled, because he was thinking 'if I go with a heroin addict, what will people think of me?' I felt ashamed and thought how bad a man I am. My brother asked me to go ahead and that he would follow rather than going along with me. Then my friend asked me in front of my brother how many books he needed and he gave a letter to one of his bookshop friends so that he could place the order. I went to the book store with the letter and they asked many questions when they saw my face, an addicted face! When they were about to hand over the book someone asked, 'will you now take the book and sell it in another bookshop?' I asked why they said this and one of them told me that I looked like an addict. Then I thought what a strange situation. Nobody believes me, even my younger brother. This is just one of the many insults that I have faced as an addict'.

Family discovery of a member's addiction gives rise to a family crisis. Family and relatives try to argue why s/he should leave addiction or make threats but typically no positive results come out. Separation or divorce may take place and then it is ordinarily not a long step towards allowing neighbours to know. Family members, relatives and neighbours gradually stop trusting the addicted person with money due to his/her image and social role. Addicts are progressively left out of important decisions regarding the family or village; their opinions carry less and less weight. They receive less emotional support from others, despite their increased need for it as their finances and health deteriorate. Although loss of identity and self-esteem is not frequently given as a reason for seeking treatment, it is often an important conscious element in the addict's maintaining abstinence after treatment.

Prejudice Towards Addicts Many addicts in Bangladesh are targeted for extortion and beatings by the police or street gangs. Most of my interviewees raised one common issue: state discrimination towards them. In Bangladesh, many of the local police still see drug users as engaged in 'social evils'. Basically, the police force them to leave public areas and abuse them physically, which often drives drug users away from life-saving care, and fuels the spread of HIV. The addicts think that the government should leave them alone and instead target the drug smugglers and dealers. They believe that if the state can control the peddlers, then the number of '*khor*' (fully addicted) will decrease. Kabir said that '*you find many drug users with broken fingers or legs due to police beatings, but those who are the sellers, making money and destroying us, they are out of reach*'. During an FGD with addicts at Jessore I was given another example of the social status of addicts:

> 'If any normal person falls or dies on the street, people will come and collect the sick person or dead body but in the case of addict, particularly a heroin 'khor', nobody will come forward to give him service even if he dies. Basically, people hate us, but we can't leave this addiction'.

Some addicts said if they want to get a job, nobody will show any interest in employing them because of their negative reputation. Biswajit said in this regard that '*many shop owners do not want to take me as staff due to my previous habit and mixing with bad friends. In my area many people know me, so I am trying to get a job elsewhere*'. When drug users become *khor*, they stop contributing to the family because they need to spend their whole income on drugs. As a result, they lose their status. However, many doctors view treating addicts as unattractive. Treating addicts is sometimes seen as a low-status occupation within psychiatry. Even physicians then practise a form of discrimination against addicts. Basically, when an addict goes to a government hospital after a beating, the doctors insist on a guardian being present before they will start treatment. The doctors' fear is that addicts are likely to have a heart problem, so they need to be cautious. But nobody wants to be the guardian of an addict.

The adverse consequences of the relationship between drug users and society are various. As society shows discrimination, so the number of marginalized addicts will increase. Addicts' ties to the social fabric decrease and are replaced by a sense of alienation and a desperation to continue to fund their drug habit. As a result, their treatment and health education options are further threatened. As a whole, society should acknowledge drug use as a social problem. The criminalization of drug users makes it very difficult to eradicate the problem from society.

3.4 Discrimination Towards PLWH

The AIDS epidemic exposes hidden vulnerabilities in the human condition, not only the individual's physical health but also a series of psychological pressures at several levels (Anderson & Bury, 1988). For most HIV-positive people, these pressures lead to what has been called 'biographical disruption' (Bury, 1982). That is to say, their sense of themselves is radically changed by the reality of their illness and also by the way they believe others would view them. People with HIV/AIDS face many problems as AIDS is a severely stigmatizing illness (Richardson, 1989). The stigma associated with HIV and AIDS can have the effect of polarizing individuals within families into those who accept the 'positive' and those who reject them (Powell-Cope & Brown, 1992). Most alarming has been the frequent discrimination against families with an HIV-infected member. Schools have revisited the admission of children with AIDS and some communities have ostracized and, on occasion, expelled such families in an effort to ensure their isolation (Bennett, 2005). Stigma and social exclusion related to HIV and AIDS have numerous effects, most notably in relation to the effect of concealment on the potential for emotional support throughout the disease trajectory (Carlisle, 2001). Downing Jr. (2008) explores the ways in which individuals living with HIV/AIDS interact with their home environments. HIV/AIDS may be illustrative of a more general phenomenon in which individual and population vulnerability to disease, disability and premature death is linked to the status of respect for human rights and dignity (Mann, Gruskin, Grodin, & Annas, 1999).

Stigma and discrimination associated with HIV/AIDS hinder the prevention of further infections, as well as the provision of care and support for PLWH. Of particular concern to HIV-infected persons and their families is the issue of access to adequate medical care and humane delivery of services (Bennett, 2005). Persons with AIDS worry that they will be denied medical care or will not receive sufficient care due to professionals' fears about the disease or attitudes regarding persons in the identified 'risk groups'. In other words, there is a tendency for medical practitioners to assume that being PLWH is a direct result of their engaging in high-risk activities, such as sexual promiscuity. In a study, Alonzo and Reynolds (1995) noted how stigma shifts during the life course after diagnosis, resulting in changes of identity and social acceptance. Muyinda et al. (1997) and Upayokin (1995) also found that HIV/AIDS-related stigma has had an adverse effect on the treatment seeking behaviours of PLWH and the coping strategies of their families. Discrimination is an overt act of unequal treatment or evaluation based on group or identity status, and may also include actions that result in unequal consequences due to identity status (Baker, 2001). Discrimination based solely on disease status has not yet received sufficient attention as a human rights violation (Annas, 1999). Elmore (2006) discussed how PLWH try to receive support from family and friends. They find also better health support from health care. A debate of 'social complexity' of ART in resource-limited environments has been discussed by Gilbert and Walker (2009). They found that AIDS patients see the ARV as 'life saving' and show the commitment and trust to health professionals.

Social prejudice may play an analogous role in the development of psychological symptoms in the AIDS patient. Prejudice distorts judgment with negativism and sensitivity; victims of discrimination are seen as fundamentally different from other persons, less deserving of basic human rights, such as the treatment they receive (Khan, 2007; Moral, 2007; Panos, 2006). Although fears of social abandonment are common in those facing a life-threatening illness, patients are immediately confronted with this reality (Zannat, 2008). In Bangladesh, like many other developing countries in the world, social stigma in respect of HIV and AIDS is extreme and extensive, particularly for PLWH. Almost all PLWH in Bangladesh try to hide their status from friends and community (Foreman, 1999; Paul, 2009). According to NASP (2016) a total of 4721 HIV cases have been identified and only 50% of identified PLWH are receiving antiretroviral therapy (ART). Here, identity problem and societal negligence along with discriminations from health care professionals towards PLWH are elaborately described with a few case studies.

PLWH's Identity and Societal Neglect Stigma and discrimination against PLWH is prevalent in Bangladesh and inhibits both the physical and mental well-being of those carrying the HIV. Both educated urban men and rural men are often secretive, as they are afraid of discrimination and of 'losing face' and status if they admit that they have the virus. Any person detected as PLWH, if their confidentiality is not kept, will face widespread and extreme social stigma. Patients soon find that neighbours consider it to be a 'bad disease'. For fear of being discriminated against at every step and also being subjected to inhuman harassment and suffering, the main concern of that person becomes to keep his disease secret as long as possible, otherwise that person may be forced to leave the village. Sometimes a patient's name and address are exposed in the media and can make headlines in the newspapers. As a result, patients become isolated from their home and family and marginalized in society. Their children cannot go to school or play with others. Under the circumstances, the person is not only denied treatment facilities to increase longevity and maintain normal health but is also forced to pose a potential health hazard for others around him, with the potential for infecting unsuspecting healthy people.

In south Asian cultures, where unorthodox sex is considered a sin, stigmatization of the patient occurs due to the mode of transmission, notably from family and close relatives as a punishment for the so-called misconduct. During my field work, I heard a lot about what could be called the 'silent torture' of PLWH, particularly in rural areas. Most cases are due to the fear, confusion and hesitation of the villagers, who have many misconceptions about HIV and they have no hesitation in making trouble for the HIV-infected person. I met with a PLWH in Khulna named Shanti who has been facing this kind of social neglect by her neighbours and other villagers for a long time:

> 'As soon as local people somehow got to know about my problem, they put pressure on my brother and blamed me for the disease. They asked him to leave the village, and threatened to evict us all. Our village is actually very conservative. I couldn't go out anywhere, and even my daughter was prevented from playing with her friends. When my brother denied the matter, some people became angry and our house was declared off limits. Nobody visits us and even poor people do not want to take food with us. If we offer anything to the local beggars they sometimes refuse. Our neighbouring families don't mix with us because they believe that if I talk to them or visit their house, they will catch the problem. They don't permit me to go to the pond and if I go to tube-well they bar me from touching it. They have the idea that if I touch the water, all of it will be contaminated with HIV. They use vulgar language against me. My mother has to bring water from the tube-well or pond for my bath, which I take in our house yard' (she was crying).

The distress associated with having a fatal, stigmatized illness can destroy the individual's sense of self or identity. People with AIDS can face considerable blame for their illness and are often deemed a group unworthy of support. These are clearly negative events that would be expected to impact on one's life to various degrees. The psychological impact of the disease varies from person to person. Although Shanti's brother could probably have stayed in the village, he would have faced many problems, social and economic, on a daily basis. Shanti herself was deeply remorseful about her stigmatized identity and faced many distressing situations. Lack of scientific knowledge about how HIV is transmitted can lead, for instance, to the idea that the virus can be contracted from burial grounds. Another reason why someone with AIDS may not mix socially is because they feel too upset or embarrassed about their physical appearance. Apart from weight loss, a person with AIDS may have disfiguring lesions on their face and other parts of the body, which are socially stigmatizing. Women are particularly accused and abused, and occasionally expelled from their marital homes. They are dispossessed of their inheritance and property rights by their families because of their infection.

The stigmatization and exclusion of HIV/AIDS patients from the community prevents them from seeking testing, counselling and treatment. There are a good number of at-risk people who do not want even to be tested for fear of social isolation. Many sufferers say disclosure of their status mainly through the media has caused their isolation. Media coverage of AIDS is sometimes highly sensational, making moral judgments between the 'innocent' and 'guilty'. Some patients criticize the role of the 'irresponsible' local media that disclosed their names or even NGOs whose activities may inadvertently identify positives to their neighbours. I had the opportunity to talk with some family members of a PLWH, Badol, brother of Shanti, in Khulna. His sister was identified in just such an insensitive way:

> 'The NGO people gathered at our house and took photographs and a video of my sister, and this affected us greatly. The local villagers became suspicious and asked us why these NGO people were visiting our house so frequently and where they came from. Some curious people went to the NGO office and saw their signboard and realized that my sister must be infected with HIV'.

Due to frequent campaigns about HIV transmission and NGO influence, some people are now more favourable in their attitude but many PLWH reported to me that villagers still have fears and confusion. As an example, Shanti, who previously had not been allowed to participate in religious devotions, recently found that this changed. Her HIV-positive daughter now plays with other children, but she still observes that people are cautious when mixing with her and her family members. In one sense, villagers have developed some sympathy for her but they still hesitate to treat her normally. Like Shanti, many PLWH remain afraid that their neighbours might find out about their HIV status because they would face serious problems, for instance, in arranging their relatives' marriages. There have been well-funded campaigns on HIV awareness but the response for testing is still very poor. Social stigma and economic constraints remain as major barriers to the treatment of AIDS patients. It seems essential from a humanitarian, and also from a social, point of view that the infected person should be diagnosed and given guidance and medication. In addition, a de-stigmatization campaign about PLWH would help.

Doctors' Discrimination In developing countries, HIV/AIDS is like a death sentence for those who do not receive treatment. The national HIV/AIDS policy in Bangladesh states that AIDS patients will be treated within the existing health care system. According to NASP documents, treatment, care and support services for PLWH have been provided through NGOs. Until recently, ART was distributed solely through NGOs. Since the end of 2012, the government has taken responsibility for procuring ART and since 2015 started providing ARVs through selected government hospitals through collaboration with NGOs under the HPNSDP. However, problems have been encountered in providing uninterrupted services because of lack of preparedness of the facilities in delivering these services as well as lacking a standard protocol for supply chain management. No health care institution or health care worker has the right to refuse to provide treatment to AIDS patients or to those with HIV infection. But almost none of the hospitals will agree to admit them if their status is known in advance. In the following, I will describe an example of attitude of doctors and nurses towards a PLWH and her non-positive family member.

Diba, a PLWH, I met in Khulna, described a difficult situation when she went to the Divisional Medical College Hospital in Khulna along with her ill brother:

> 'Once my elder brother got sick and I took him to the Medical College Hospital for treatment. Some of the doctors knew me to be 'positive', so rumours spread that my brother was also suffering from HIV. I felt very sad and insulted when I saw the Specialists, who regarded me as an object of mockery. They visited the other patients but not to my brother. When the doctors claimed my brother was PLWH, I had his blood tested and he was found to be 'negative'. Despite this, still no doctor saw him and I heard the nurses also whispering about the issue. I asked them why they thought he was 'positive'. Was it because of his fever? Were there no other symptoms of HIV? I tried to challenge their prejudice but failed, and then I took my brother to a private clinic. Next day there was news in a local newspaper that a PLWH had been found and that he had left the hospital due to fear. Even when I was in the government infectious disease centre with my husband in Dhaka, the nurses didn't give me the medicine themselves. They wrapped the medicine in paper and left it outside the room'.

This is a common reality that most PLWH face outright discrimination by medical service providers, including doctors, both in government hospitals and private clinics in Bangladesh. Many NGO programme officers in the field reported how nurses have misconceptions about the cause and spread of HIV. Some believe that HIV can spread on the breath or through the saliva when spitting. This is due to poor training. Although the number of PLWH is increasing, the only treatment option is the Infectious Diseases Hospital (IDH) in Dhaka and a few private specialized hospitals. The state-run IDH has provided treatment guidelines for HIV/AIDS patients since 1989. But they do not get proper treatment as the IDH is yet to provide antiretroviral (ARV) therapy for these patients since it is very expensive. Although antiretroviral drugs (ARVs) have been registered for manufacture in Bangladesh, they are not available through the public health care system. The IDH can provide only around 50 percent of drugs required for AIDS. It just provides treatment for opportunistic infections that occur due to the collapse of the patients' immune systems. Ideally, they also need psychological and nutritional support. The IDH has limited facilities and still there is staff prejudice. There are stories of nurses and other medical service providers refusing to change bedpans, feed, wash or even talk to someone in their care who has AIDS. There was a newspaper report that a nurse refused to attach an oxygen mask to a dying patient and instead sent a cleaner. Sometimes, in other hospitals, medical service providers force the patient silently to leave the hospital bed when they come to know about their disease (personal communication with many NGO officials in field). In an interview, one HIV expert told that during a visit to a hospital, he found the doctors and nurses on duty were not giving intravenous saline to a PLWH patient, though there was express advice for that. These are the expressions of serious violations of human rights towards PLWH. Most unfortunately, in many cases, PLWH relatives also face discrimination because there are doctors who know the family's circumstances and who then refuse treatment to them all (personal communication with many NGO officials in the field) like Diva and her brother. Known PLWH are turned away from government hospitals and they are forced to fall back on private clinics, but even here they cannot stay long. Due to a lack of acceptance of this disease among the doctors and nurses, AIDS patients present with false ailments. When the clinic doctors come to know of the patient's status from their previous prescription, it is common for this information to spread to the whole clinic and, as a result, the patient has to change clinic because of security concerns. Confidentiality of a person's HIV antibody status is particularly important because of the potential personal, social and economic harm that may result from disclosure of this information.

In Bangladesh, PLWH also fear journalists because local papers have no compunction about publishing their medical details, causing distress and trouble for their friends and family. Although Diba eventually found treatment for her brother, she considered herself to have been discriminated against and victimized. She also felt insulted and sad about the negligence of the doctors in the medical college hospital. One retired justice of South Africa, Edwin Cameron, questioned in a UN forum 'when any identified PLWH comes to know that he/she will not get treatment, do we have the right to ostracize him/her after that?' PLWH are stigmatized, isolated and deprived of care, despite the fact that the UN Commission on Human Rights has confirmed that discrimination against PLWH, or those thought to be infected, is a clear violation of their human rights. This declaration has been signed and ratified by all member countries including Bangladesh. There is some hope because, although in the early years of the epidemic people affected by HIV were discriminated against by neighbours, friends and sometimes family members, over the years attitudes have been changing gradually.

3.5 Identity, Stigma and HIV Danger

In Bangladesh, gender-based violence, stigma and discrimination, and certain laws are hindering the HIV response. The social situations of many marginalized people like sex workers, drug users and PLWH have exposed them to violence and stigma due to their negative identities in society. This has crushed their self-esteem. Violence is a significant threat to marginalized people's physical and psychological well-being, especially for poor, marginalized people who are not well-linked to health and social services. Stigma has resulted in widespread discrimination against those marginalized people. For example, Bangladeshi society discriminates against sex workers as immoral women. Although violence is not a major problem at different brothels in Bangladesh, sex workers are heavily stigmatized by their family, relatives and neighbouring community. Second, uncontrolled use of narcotic drugs is apt to evoke social condemnation rather than sympathy, and more likely to result in legal punishment than medical care. All drug users are stigmatized but heroin addicts are often regarded with particular abhorrence. Third, PLWH have to face cruelty and social boycotts because of superstitions, myths and misconceptions about HIV infection. Sex workers, drug users and PLWH often refrain from seeking medical treatment because of the risk that they will be identified as sex workers, or drug users or PLWH within their community and ostracized.

In my fieldwork, I found that stigma blocks the way for marginalized people to access appropriate medical health care and leads to discrimination by families or communities. Many of them are denied property and inheritance rights. Because of the increasing spread of HIV/AIDS in vulnerable communities, identity and stigma are interconnected, which feeds prejudices and fuels victimization. A rights-based, participatory approach, comparatively new in the field of public health, is required not just for tackling the HIV epidemic but for promoting health and better quality of life among CSWs, DUs and PLWH because their life and occupation is deeply entrenched in gender discrimination, exploitation and marginalization. Human rights and gender equality should be the focus of the national HIV response. Returning to the central thesis of this chapter, there is the tendency to regard HIV/AIDS as a problem to be dealt with at the bio-medical/behavioural level or as an individual human rights issue rather than seeing it as a broad social and developmental issue. In the next chapter, risk behaviours, consciousness and risk coping issues of the marginalized and vulnerable people will be discussed.

References

Alexander, P. (1996). Foreword. In N. Mckeganey & M. Barnard (Eds.), *Sex work on the streets: Prostitutes and their clients*. London, UK: Open University Press.

Alonzo, A. A., & Reynolds, N. R. (1995). Stigma, HIV and AIDS: An exploration and elaboration of a stigma trajectory. *Social Science and Medicine, 41*(3), 303–315.

Anderson, R., & Bury, M. (Eds.). (1988). *Living with chronic illness: The experience of patients and their families*. London, UK: Hyman Unwin.

Annas, G. J. (1999). The impact of health policies on human rights: AIDS and TB control. In J. M. Mann, S. Gruskin, M. A. Grodin, & G. J. Annas (Eds.), *Health and human rights*. New York, NY: Routledge.

Ara, Z. (2005). Violation & denial of access to health-rights for women involved in commercial sex work in Bangladesh. *Women's Health and Urban life, 4*(1), 6–26.

Asthana, S. (1996). AIDS-related policies, legislation and programme implementation in India. *Health Policy and Planning, 11*(2), 184–197.

Baker, N. L. (2001). Prejudice. In J. Worell (Ed.), *Encyclopedia of women and gender* (Vol. 2). New York, NY: Academic Press.

Barnard, M. (2007). *Drug addiction and families*. London, UK: Jessica Kingsley Publishers.

Benjamin, C. (2001). Aspects of stigma associated with genetic conditions. In T. Mason, C. Carlisle, C. Watkins, & E. Whitehead (Eds.), *Stigma and social exclusion in healthcare*. London, UK: Routledge.

Bennett, A. (2005). *Culture and everyday life*. London, UK: Sage.

Blacking, J. (1983). The concept of identity and folk concepts of self: A Venda case study. In A. Jacobson-Widding (Ed.), *Identity: Personal and socio-cultural, A symposium*. Stockholm, Sweden: Uppsala, Academiae Upsaliensis.

Blanchet, T. (1996). *Lost innocence: Stolen childhood*. Dhaka, Bangladesh: University Press Limited (UPL).

Blithe, S. J., Wolfe, A. W., & Mohr, B. (2019). *Sex and Stigma: Stories of everyday life in Nevada's legal brothels*. New York, NY: New York University Press.

Bungay, V., Halpin, M., Atchison, C., & Johnston, C. (2011). Structure and agency: Reflections from an exploratory study of Vancouver indoor sex workers. *Culture, Health & Sexuality, 13*(1), 15–29.

Bury, M. (1982). Chronic illness as biographical disruption. *Sociology of Health and Illness, 4*(2), 167–182.

Butler, J. (2004). *Undoing gender*. New York, NY: Routledge.

Carlisle, C. (2001). HIV and AIDS. In T. Mason, C. Carlisle, C. Watkins, & E. Whitehead (Eds.), *Stigma and social exclusion in healthcare*. London, UK: Routledge.

Carnwath, T., & Smith, I. (2002). *Heroin century*. London, UK: Routledge.

Chant, S., & Mcllwaine, C. (1995). *Women of a lesser cost: Female labour, foreign exchange, and Philippine development*. London, UK: Pluto Press.

Chowdhury, R. (2006). "Outsiders" and identity reconstruction in the sex workers' movement in Bangladesh. *Sociological Spectrum, 26*, 335–357.

Clarke, A. (2001). *The Sociology of health care*. Harlow, UK: Pearson.

Craddock, S. (2012). Drug partnerships and global practices. *Health and Place, 18*(3), 481–489.

Crocker, J., Major, B., & Steele, C. (1998). Social stigma. In D. T. Gilbert, S. T. Fiske, & G. Lindzey (Eds.), *The handbook of social psychology* (Vol. 2, 4th ed., pp. 504–553). New York, NY: McGraw-Hill.

Currie, E. (1993). *Reckoning: Drugs, the cities and the American future*. New York, NY: Hill and Wang.

Dorn, N., & South, N. (Eds.). (1987). *A land fit for heroin? Drug policies, prevention and practice*. London, NY: Macmillan Education.

Downing Jr., M. J. (2008). The role of home in HIV/AIDS: A visual approach to understanding human-environment interactions in the context of long-term illness. *Health & Place, 14*(2), 313–322.

Elmore, K. (2006). The migratory experiences of people with HIV/AIDS (PWHA) in Wilmington, North Carolina. *Health and Place, 12*(4), 570–579.

Erikson, E. H. (1968). *Identity: Youth and crisis*. New York, NY: Norton.

Fassi, M. N. (2011). Dealing with the margins of law: Adult sex workers' resistance in everyday life. *Oñati Socio-Legal Series, 1*(1), 36.

Fischhoff, B. (2001). Defining stigma. In J. Elynn, P. Slovic, & H. Kunreuther (Eds.), *Risk, media and stigma: Understanding public challenges to modern science and technology*. London, UK: Earthscan.

Foreman, M. (1999). *AIDS and men: Taking risks or taking responsibilities?* London, UK: The Panos Institute, Zed Books.

Franzoi, S. L. (1996). *Social psychology*. London, UK: Brown & Benchmark.

Ghosh, J., Wadhwa, V., & Kalipeni, E. (2009). Vulnerability to HIV/AIDS among women of reproductive age in the slums of Delhi and Hyderabad, India. *Social Science and Medicine, 68*(4), 638–642.

Gilbert, L., & Walker, L. (2009). "They (ARVs) are my life, without them I'm nothing"-experiences of patients attending a HIV/AIDS clinic in Johannesburg, South Africa. *Health and Place, 15*(4), 1123–1129.

Goddard, V. (1993). Honour and shame: The control of women's sexuality and group identity in Naples. In P. Caplan (Ed.), *The cultural construction of sexuality*. London, UK: Routledge.

Goffman, E. (1963). *Stigma: notes on the management of spoiled identity*. Upper Saddle River, NJ: Prentice-Hall.

Gutierrez-Garza, A.. (2013). *The everyday moralities of migrant women: life and labour of Latin American domestic and sex workers in London.* PhD thesis, London, UK: The London School of Economics and Political Science (LSE).

Hogg, M. A. (1985). Masculine and feminine speech in dyads and groups: A study of speech style and gender salience. *Journal of Language and Social Psychology, 4*, 99–112.

Hogg, M. A., & Vaughan, G. M. (2002). *Social psychology* (3rd ed.). Essex, UK: Pearson Education.

Holzner, B., & Robertson, R. (1980). Identity and authority: A problem analysis of processes of identification and authorization. In R. Robertson & B. Holzner (Eds.), *Identity and authority: Explorations in the theory of society*. Oxford, UK: Basil Blackwell.

Hopper, S. (1981). Diabetes as a stigmatizing condition. *Social Science and Medicine, 15B*, 11–19.

Jenkins, R. (1999). *Social identity*. London, UK: Routledge.

Kasperson, R., Golding, D., & Kasperson, J. X. (2005). Stigma and the social amplification of risk: Towards a framework of analysis. In J. X. Kasperson & R. E. Kasperson (Eds.), *The social contours of risk* (Vol. 1). London, UK: Earthscan.

Katz, I. (1981). *Stigma: A social psychological analysis*. Hillsdale, NJ: Lawrence.

Khan, A. W. (2007). Experts worried at increasing HIV cases among IDUs. Conference on AIDS, *The Daily Star*, August 24, Dhaka.

Khan, S. (1988). *The fifty percent: Women in development and policy in Bangladesh*. Dhaka, Bangladesh: University Press Limited (UPL).

Khosla, N. (2009). HIV/AIDS interventions in Bangladesh: What can application of a social exclusion framework tell us? *Journal of Health, Population and Nutrition, 27*(4), 587–597.

Knight, V. C. (2006). *Drug users at risk to HIV: Documenting our experience 2000–2005, HIV program*. Dhaka, Bangladesh: Care Bangladesh.

Lawler, S. (2008). *Identity: Sociological perspectives*. Cambridge, UK: Polity Press.

Lewis, J. A., Dana, R. Q., & Blevins, G. A. (2002). *Substance abuse counseling*. Boston, MA: Thomson Learning.

Lex, B. W. (2000). Gender and cultural influences on substance abuse. In R. M. Eisler & M. Hersen (Eds.), *Handbook of gender, culture and health* (pp. 255–297). Hillsdale, NJ: Lawrence Publishers.

Mann, J. M., Gruskin, S., Grodin, M. A., & Annas, G. J. (Eds.). (1999). *Health and human rights: A reader*. New York, NY: Routledge.

Mason, T., Carlisle, C., Watkins, C., & Whitehead, E. (Eds.). (2001). *Stigma and social exclusion in healthcare*. London, UK: Routledge.

McDermott, P. (1992). Representations of drug users facts, myths and their role in harm reduction strategy. In P. A. O'Hare, R. Newcombe, A. Matthews, E. C. Buning, & E. Drucker (Eds.), *The reduction of drug-related harm* (pp. 195–201). London, UK: Routledge.

Mercer, K. (1998). Welcome to the jungle: Identity and diversity in postmodern politics. In J. Rutherford (Ed.), *Identity: Community, culture, difference*. London, UK: Lawrence and Wishart.

Miller, R. (1994). What drugs do to users. In R. Coomber (Ed.), *Drugs and drug use in society: A critical reader* (pp. 5–23). Kent, UK: Greenwich University Press.

Moral, S. (2007). Notes from AIDS conference, on the world AIDS day, The Prothom Alo (*Bangla daily*), December 1, Dhaka.

Muyinda, H., Seeley, J., Pickering, H., & Barton, T. (1997). Social aspects of AIDS-related stigma in rural Uganda. *Health & Place, 3*(3), 143–147.

NASP. (2016). *Fourth National Strategic Plan for HIV and AIDS response, National AIDS/STD Program*. Dhaka, Bangladesh: Ministry of Health and Family Welfare, Government of Bangladesh.

O'Neill, M. (1997). Prostitute women now. In G. Scambler & A. Scambler (Eds.), *Rethinking prostitution: Purchasing sex in the 1990s*. London, UK: Routledge.

Panos. (2006). *Keeping the promise? A study of progress made in implementing the UNGASS declaration of commitment on HIV/AIDS in Bangladesh*. Dhaka, Bangladesh: The Panos Global AIDS Programme.

Parker, R., & Aggleton, P. (2003). HIV and AIDS-related stigma and discrimination: A conceptual framework and implications for action. *Social Science and Medicine, 57*(1), 13–24.
Patterson, C. B. (2015). Beyond the stigma: The Asian sex worker as First World savior. In M. Laing, K. Pilcher, & N. Smith (Eds.), *Queer sex work* (pp. 53–65). Oxon, UK: Routledge.
Paul, A. (2009). *Geographies of HIV/AIDS in Bangladesh: Vulnerability, stigma and place*. Durham theses, Durham University. http://etheses.dur.ac.uk/1348/
Phoenix, J. (2000). Prostitute identities: Men, money and violence. *British Journal of Criminology, 40*(1), 37–55.
Powell-Cope, G. M., & Brown, M. A. (1992). Going public as an AIDS family caregiver. *Social Science and Medicine, 34*(5), 571–580.
Ransome, Y., Kawachi, I., Braunstein, S., & Nash, D. (2016). Structural inequalities drive late HIV diagnosis: The role of black racial concentration, income inequality, socioeconomic deprivation, and HIV testing. *Health and Place, 42*(November), 148–158.
Rawat, A., & Kumar, S. (2015). Lifeworlds of children of sex workers in Budhwar peth. *The Indian Journal of Social Work, 76*(4), 579–596.
Richardson, D. (1989). *Women and the AIDS crisis*. London, UK: Pandora Press.
Robertson, R. (1987). *Heroin, AIDS and society*. London, UK: Hodder and Stoughton.
Scambler, G. (2020). Towards a sociology of shaming and blaming. In G. Scambler (Ed.), *A sociology of shame and blame* (pp. 87–103). Cham, Switzerland: Springer, Palgrave Pivot.
Singer, M. (2001). Toward a bio-cultural and political economic integration of alcohol, tobacco and drug studies in the coming century. *Social Science and Medicine, 53*(2), 199–213.
Sontag, S. (1989). A new book on stigma. In S. Rogers (Ed.), *Explaining health and illness: An exploration of diversity*. New York, NY: Harvester Wheatsheaf.
Sultana, H. (2015). Sex worker activism, feminist discourse and HIV in Bangladesh. *Culture, Health & Sexuality, 17*, 6777–6788.
Ulasi, C. I., Preko, P. O., Baidoo, J. A., Bayard, B., Ehiri, J. E., Jolly, C. M., et al. (2009). HIV/AIDS-related stigma in Kumasi, Ghana. *Health and Place, 15*(1), 255–262.
Ullah, A. K. M. A. (2011). HIV/AIDS-related stigma and discrimination: A study of health care providers in Bangladesh. *Journal of the International Association of Physicians in AIDS Care, 10*(2), 97–104.
Upayokin, P. (1995) Treatment choice, disease outcome and stigma: An investigation of leprosy patients and illness behaviour in Thailand. U-M-I Dissertation Services, Michigan.
Usher, K., Jackson, D., & O'Brien, L. (2005). Adolescent drug abuse: Helping families survive. *International Journal of Mental health Nursing, 14*(3), 209–214.
van Blerk, L. (2016). Livelihoods as relational Im/mobilities: Exploring the everyday practices of young female sex workers in Ethiopia. *Annals of the American Association of Geographers, 106*(2), 413–421.
Walker, R. (2017). Selling sex, mothering and 'keeping well' in the city: Reflecting on the everyday experiences of cross-border migrant women who sell sex in Johannesburg. *Urban Forum, 28*, 59–73.
Ware, N. C., Wyatt, M. A., & Tugenberg, T. (2005). Adherence, stereotyping and unequal HIV treatment for active users of illegal drugs. *Social Science and Medicine, 61*(3), 565–576.
Williams, B. (1999). *Working with victims of crime: Policies, politics and practice*. London, UK: Jessica Kingsley Publishers.
Wilton, T. (1994). Feminism and the erotics of health promotion. In L. Doyal, J. Naidoo, & T. Wilton (Eds.), *AIDS: Setting a feminist agenda*. London, UK: Taylor & Francis.
Zannat, M. (2008). AIDS patients deprived of proper care, *The Daily Star*, April 18, Dhaka.
Zierler, S. (1997). Hitting hard: HIV and violence. In N. Goldstein & J. L. Manlowe (Eds.), *The gender politics of HIV/AIDS in women: Perspectives on the pandemic in the United States*. New York, NY: New York University Press.

Chapter 4
HIV Risk Behaviour, Consciousness and 'Risk Coping'

4.1 Introduction

Geographers have worked on infectious diseases such as cholera, malaria, influenza, measles and hepatitis for a long time (for example, Cliff & Haggett, 1988; Learmonth, 1952; May, 1958; Pyle, 1969; Stamp, 1964; Thomas, 1992) and are now making contributions to the geography of HIV/AIDS in the contemporary period. From the early years of the HIV/AIDS epidemic, it was apparent that international travel—for business and for pleasure—played a necessary, though insufficient, role in the geographical and social diffusion of HIV infection (Gould, 1993; Hawkes, 1992). HIV is transmitted from person to person primarily by unprotected sex or through the sharing of injecting equipment. Discourses around health care, sexuality, gender and migration related to the epidemiology of HIV/AIDS, and local processes involving individual and community response are much argued about by geographers, anthropologists and others. These concepts are also related to class, gender, sexuality, culture and politics (Brown, 1995; Patton, 1994; Takahashi & Dear, 1997). Geographers have given their attention to the virus itself, developing diffusion models or recounting origins theories without reference to the spatial-political implications of living with AIDS (Kearns, 1996). Many HIV/AIDS-based studies have shown people's lifestyle changes, high-risk behaviours and health care needs (Ellis & Muschkin, 1996). Geographers have also highlighted risk theory, transmission modes, sexuality and 'new waves' of risk and vulnerability in the spread of the HIV epidemic (Kalipeni, Craddock, Oppong, & Ghosh, 2004; Moran, 2005; Teye, 2005). These studies have drawn attention to the impact of HIV on individual behaviour and the conditions of life from an ecological perspective.

However, few recent theoretical works have tried to connect the social and behavioural mechanisms (for example, Cheng, 2005; Lindquist, 2005; Marten, 2005) which play a role in shaping the geography of new HIV infections. All of the quantitative-based works have determined that HIV/AIDS has tended to cluster in

A. Paul, *HIV/AIDS in Bangladesh*, Global Perspectives on Health Geography,
https://doi.org/10.1007/978-3-030-57650-9_4

certain areas and infection, diffusion and overlap can be expected between different population subgroups, even those who have not traditionally been at risk. In some recent discourses, much empirical research (Cianelli et al., 2013; Gagnon, Merry, Bocking, Rosenberg, & Oxman-Martinez, 2010; Kuhanen, 2010; Madise et al., 2012) shows a connection between health risk behaviours and HIV knowledge among the hard-to-reach people. In their discussion, how behavioural risk can affect knowledge, attitudes and practices towards HIV and STD is illustrated. Moreover, despite a growth of studies in different areas addressing questions of risk and its social context, the dominant paradigm of 'risk groups' and individual behaviours continues to be a key focus (Craddock, 2000). The concepts of 'risk group' and 'risk behaviour' are well established in public health research as well as by geographers to identify the determinants of healthy lifestyles as well as chronic ill-health-related results (Shaw, Dorling, & Mitchell, 2002) including behaviour (e.g. for HIV/AIDS, unprotected sex). The nature and intensity of risk varies by gender, race, ethnicity and class and, in addition, the risk group idea moves in the direction of peopling the virus (Kearns, 1996). In this way, geographers have been constructing 'new geographies of HIV/AIDS', which are alternatives to the epidemiological approach.

Public health 'problems' involve not only concern for the exposure of populations to bio-medical risks, but also concern for managing social risks such as fear, apathy and misinformation (Lawrence, Kearns, Park, Bryder, & Worth, 2008). Contemporary constructions of HIV risk in epidemiological research associate unprotected sex and needle sharing as 'risk behaviours' among the 'most at-risk' groups. Without having a gateway to health knowledge and self-protection, sex workers and drug users are very much susceptible to health risks, particularly HIV infection. Here, individual perceptions of risk susceptibility are influenced by everyday understandings of 'risk acceptability' (Rhodes & Quirk, 1996). Risk perceptions are socially constructed, and individual behaviours are driven by perceptions or beliefs about risks (Frewer, 1999). In recent debate, a discussion of risk behaviours among sex workers, Marshall et al. (2009) examined the role of environmental factors in transmission of HIV and STI risk among customers of sex workers. Kuhanen (2010) argues that the AIDS outbreak in Uganda was a consequence of economic change, erosion of conventional sexuality, sexualized spaces and sexual networking. Goldenberg et al. (2011) discussed the role of high payment for unprotected sex which ultimately increases the health risk among both sex workers and their clients. Regarding drug users, Tempalski and McQuie (2009) discussed the drug user's cultural norms, stigma and ecological factors which may affect risk behaviours. Werb et al. (2010) compared drug-related risks and behaviours among street youths residing in two adjacent neighbourhoods. Parkin and Coomber (2011) found some settings which contribute to a wide range of injecting-related harm and hazard. In addition, Luginaah (2008) showed a relationship between drug use behaviours among the transport workers and women's exposure to HIV or STD due to poverty and sexual violence.

Bangladesh is currently a low HIV prevalence country with its own at-risk populations (FHI, 2006; Jenkins & Rahman, 2002), but the threat of an HIV epidemic is looming over the country, as STIs and other indications of risky behaviour have been found to be high. Recent Behavioural Surveillance Survey (BSS) data indicate that the drug user population is well integrated into the surrounding urban community, socially and sexually, thus raising concerns about the spread of HIV infection (Islam et al., 2015). As a significant part of vulnerable community, truckers' risk of acquiring sexually transmitted infections is increased by having sexual relations with CSWs (Alam et al., 2007; Gibney et al., 2002), as shown in the literature on Bangladeshi sex workers' STDs (Azim et al., 2004; Choudhury, Rahman, & Moniruzzaman, 1989; Hossain, Mani, Sidik, Shahar, & Islam, 2014). There are some other groups such as garment workers, transport workers, refugees, displaced persons and some minority ethnic populations among whom higher vulnerability is suspected but supporting evidence is not strong (NASP, 2016). In order to control the spread of HIV infection, primary prevention, such as through awareness and changing behaviours, is the highest priority in HIV control programmes around the world. But awareness about HIV/AIDS among the general population of Bangladesh is poor (Asaduzzaman, Higuchi, Sarker, & Hamajima, 2016; Choudhary & Rahman, 2014; Haque, Hossain, Chowdhury, & Uddin, 2018; Mondal, Islam, Rahman, Rahman, & Hoque, 2012; Sheikh, Uddin, & Khan, 2017; Yaya, Bishwajit, Danhoundo, Shah, & Ekholuenetale, 2016). In this chapter, I will investigate the risk behaviours of vulnerable people and their level of knowledge about health risks. In addition, different risk-minimizing techniques adopted by these vulnerable people will also be addressed.

4.2 Risk Behaviours of Vulnerable People

Sex workers were incorporated into the western AIDS discourse initially as a 'risk group' category (Murray & Robinson, 1996). Female sex workers play an important role in heterosexual transmission of HIV (Gangopadhyay, Chanda, Sarkar, Niyogi, et al., 2005) because they are considered both a core group for the acquisition and transmission of STIs and HIV, and as a bridging group to the general population. Among brothel-based sex workers, Sarkar, Islam, Durandin, et al. (1998) and Nessa et al. (2005), Nessa, Waris, Sultan, Monira, et al., 2004) found a high prevalence of STDs such as syphilis, chlamydia and/or gonorrhoea, which indicate that if HIV is introduced, it will rapidly increase among brothel girls first and then among the general population. Gazi et al. (2009) found that hotel-based sex workers have a higher client turnover than their peers on the streets and in brothels. In a study of street-based sex workers in Dhaka, Motiur Rahman, Alam, Nessa, Hossain, et al. (2000) and Khanam et al. (2017) found that the prevalence of STIs among street girls was also high. Azim et al. (2006) show that infection rate of STDs, particularly syphilis, among floating sex workers is higher than brothel-based sex workers. In Bangladesh, the level of commercial sex is higher than elsewhere in Asia and condom use during commercial sex is still low (Islam & Conigrave, 2008). In 2016, consistent condom use among the female sex workers was 39.4%, 36.9% and 42.5% in brothels, streets and hotels of Dhaka, respectively (NASP, 2016). Apart from the

STD issue, many sex workers are reported to have sexual intercourse during their menstruation period (Sarkar et al., 2005) and an increasing rate of anal intercourse (Jenkins & Rahman, 2002). However, In Bangladesh, poverty and conservativeness put women in a vulnerable position (Bagley, Kadri, Shahnaz, Simkhada, & King, 2017).

Injecting drug users have traditionally been a group at high risk from early mortality and the AIDS epidemic has increased this risk. The problem of drug abuse has reached recognizably significant proportions today in Bangladesh (Islam, 2006; Kabir, Goh, Kamal, Khan, & Kazembe, 2013; Morris & Aftab, 2012). Illegal opiate use behaviour, which is considered a lifestyle risk factor for HIV, is prevalent. Repeated rounds of surveillance have revealed that the rate of sero-positivity is highest among intravenous drug users (IVDUs) and the findings also confirm the presence of high levels of behavioural risk factors for the acquisition of HIV infection through needle sharing (Azim et al., 2006; Khosla, 2009; Mondal, Takaku, Ohkusa, Sugawara, & Okabe, 2009). Injecting drug use has steadily gained in popularity in Bangladesh (Larance et al., 2011). Moreover, most of the STD/HIV-related global literature on truckers, particularly long-distance drivers (Bwayo et al., 1994; Mbugua et al., 1995) clearly shows their attachment to high-risk behaviour, particularly sexual activities which ultimately expose them to STDs and HIV (George et al., 1997; Lacerda et al., 1997; Singh & Malaviya, 1994). One study found that women living close to a truck stand in Dhaka city are likely to be at higher risk for STDs and genital tract infections (Gibney et al., 2001; Gibney et al., 2002; Sabin, 1998). Although premarital sex is not openly accepted in conservative Bangladeshi society (Caldwell, Pieris, Barkat-e-Khuda, Caldwell, & Caldwell, 1999), it is assessed in one study (Gibney, Saquib, & Metzger, 2003) that 73% of married and 78.4% of unmarried truckers were involved in sex on their travels and the mean age at first sexual intercourse was 17.8 years. Photo Series 4.1 shows some risk practice, environment and vulnerable people as symbolic. The following sections consider issues of vulnerable people's, such as commercial sex workers, drug users and truck drivers, risk behaviours individually.

Health Risk Behaviour of Sex Workers There is a concern that HIV or STD infections will increase among sex workers and their clients who will then form a 'bridge' to clients' wives and girlfriends in the general population. In terms of condom use, many girls tried to skirt around the issue when I asked. From my experience, I can say that none of them told me willingly about sex without a condom. When I asked a primary question for condom use, they would just say 'no condom, no sex' but deeper enquiry revealed that sex without a condom was possible for 'very close friends'. After spending more time with me, some girls found me trustworthy, and confessed that they need to do sex without a '*packet*' (condom) because of the money. Particularly when I asked some questions about contraceptives and pregnancy, I usually got some hints about their sexual habits and preferences. During my field work, I found substantial differences in condom use between the different types of sex workers. Many are willing to work without condoms for extra cash and these girls also have unprotected sex with their lovers or husbands.

Photo Series 4.1 Risk practice, environment and vulnerable people

Although many of them understand the risk, they cannot acknowledge it to their friends or husbands for reasons of stigma. Many non-brothel sex workers have 'good friends' who do not like to use a condom either and, if asked, would be suspicious about her character. One told me that '*I always carry a condom with me in my bag. But I don't offer it to my good friend because he might wonder why I am carrying a condom and perhaps think that I am having sex with others. Then I will have a problem*' (laughs). Among street girls, condom use depends on whether customers are known or unknown. One street girl, Subarna said '*I know many people who have been coming to me for 7–8 years. As I know them I can do 'work' without condom, but with those I don't know, I insist on a condom*'. I found some older street girls who usually do not ask for condom use as a deliberate strategy because customers will otherwise go to younger girls. However, I observed that there are some customers who are considered as regulars in the brothel who want to stay for a long time with their sex partners. Most brothel girls do not insist on condom use by their lover or *babu*. I found one girl in the Jessore brothel who confirmed this: '*He (babu) is my own man, why should he use a condom? It's not right! I want to say that I don't insist on a condom because I love him*'.

Condoms are offered by the girls depending on the type of sexual partner. For new clients, CSWs are more persuasive about condom use than with repeat or regular clients. Closer interactions revealed that many sex workers allowed their clients, as well as those forced on them by their commercial sex partners, to have sex without a condom. Condom negotiation between girls and customers makes for many chaotic situations. Customers say that, as they are paying, they will decide whether to use a condom or not. Girls think that if they argue with a customer over this issue, he will not be lost just for one day, but forever. One consideration is that some customers suggest that if they do wear a condom it will take them longer to ejaculate. Crudely speaking, sex workers want their customers to perform quickly and leave, reducing the physical impact upon themselves and making way for a new customer. Many sex workers do not persuade or try to convince their clients to use a condom because they consider it as a waste of time. In this study, women expressed serious concern that they would lose the opportunity to earn money if they ever raised the issue of condom use with their clients. Women are also afraid to insist upon condom usage because they fear violence from clients, especially from drunken and uneducated clients who do not believe that it is possible to enjoy sex with a condom. They are least bothered about protected sex and transmission of STIs and HIV. Professional immaturity of younger sex workers who may have joined the profession recently is another factor in incidences of unprotected sex with their clients, as is poverty and high levels of competition among sex workers.

Health risks associated with sex practices are an ongoing hazard associated with sex work. Most sex workers simultaneously have a number of symptoms of STDs and RTIs, such as vaginal discharge, cervical discharge, swelling over the groin, burning sensation during urination or painful recurring superficial ulcers on the vulva. Sex workers' vulnerability and risk of contracting STDs increase greatly for those unable to practise safer sex where money and timing play a role. The prevalence of STDs in Bangladesh indicates that high-risk sexual activity is occurring through extramarital and premarital sexual intercourse with sex workers. As a result, health risks, particularly of STDs, are transferred from the customer to sex worker and vice versa. In Khulna, I found one girl, Lima who told me about this risk-spreading cycle. Many times she has had sex in hotels without a condom. As she has some sexual diseases and a fear of an unwanted pregnancy, she is habituated to taking homeopathic and herbal medicines. Although she has a husband, she is not recognized by his family because he already has another wife. She said:

> 'Sometimes I feel that I am at risk because most of my work is without a condom case and I also feel that I may be transferring disease to others. For example, I am working with customers without a condom, then I am sleeping with my husband. Then he is having sex with his first wife. So, all of us are at risk'.

Similarly, I found a Benapole girl, Johura who works as a cook in the bachelor's quarters in the customs area and is involved in commercial sex work at night. The bachelors never use a condom because they say that they have no other sexual relations and so do not need to. But they do not realize that she is earning more income for her livelihood with the many transport workers at Benapole port. My conversation with her illustrated many risk dimensions of human psychology and of hypocrisy. Other than STD issue, many sex workers are reported to have sexual intercourse during their menstruation period with their partners. These types of sex practices during menstruation are also associated with a higher transmission of HIV. However, the marked increase in the proportion of women agreeing to anal intercourse is striking. Anal sex is associated with a higher risk of HIV transmission than peno-vaginal sex. It is found in both the review and 'field', female sex workers in Bangladesh frequently state that they dislike anal penetration, but the threat of diminished income is likely to have pressured their acquiescence which is also found in my study.

Although Bangladesh has a conservative society sexually, women become 'lost' due to their economic situation. Many women told me that girls become 'lost' due to their environment and poverty. Recently, in Bangladesh, access to multiple partners has become a matter of 'masculinity'. An example is that hotel customers want a different girl on each visit. Some want new, young girls, though this depends on the customer's psychology. However, one NGO programme manager in Khulna told me that there are many casual sex workers who have a husband. These are 'unreachable' for NGO workers because they work during the day as maids, factory workers, construction assistants or cleaning girls and also have multiple sexual partners. The NGO survey found that they are forced to have sex with their contractors, or mill owners in order to get a job and keep it. These girls feel cheated that they are paid nothing for this, so they decide to work independently. During my field work, I also talked with some outreach workers who are raising awareness among sex workers about HIV and are also providing condoms and other materials. One of them, Mala told me that they cannot approach many of the casual girls who use hotels because of their irregular commuting behaviour. She assumes that some girls come from the village and some are local college students. Both groups do not stay in the hotel for long and so the NGO workers find it difficult to get across the message about risky behaviour. However, many residence-based girls told me that in every residential area there are many madams' homes which have a 'good environment' or secure status. All of these 'unreachable' girls are potential sources of risk bridging. Moreover, brothel and non-brothel girls claim that diversified types of customers come to them like drivers, policemen, rickshaw pullers, officers, college and school students, civil servants and others, both married and unmarried. Married men come when their wife becomes pregnant. Old customers encourage their friends to become new customers. This customer variety can be said to enhance the risk profile of the brothel and non-brothel settings. Almost all are playing a role in disseminating risk knowingly. The girls know that they have many customers who do not use condoms and the customers insist on a high turnover of girls and this must increase the probability of a transfer of the virus.

Addicts' Realization of Risk 'Transfer' In the context of Bangladesh, although syringes are cheap and easily available in the open market, needle sharing is extremely common, and nothing is done to sterilize the equipment between uses. As a consequence, injecting drug users present a tremendous potential for an HIV epidemic due to their needle-sharing habits, while non-injecting drug users are also prone to spread or receive HIV infection through their unsafe sexual behaviour. Most of the drug addicts, particularly heroin users, look for sexual 'pleasure', and such risk takers provide a risk bridge for the transfer of sexual and other infectious diseases. I found one drug user, Shapon at Jessore, who is involved in many risks. He injects heroin and other drugs, sometimes shares needles, and is a professional blood donor to raise money for drugs. He then visits brothels or uses street girls, both without using a condom. He has some concerns about selling his blood and knows that there is no screening for whether it is pure or infected. He told me:

> 'I always feel that I am doing wrong about drug addiction, sex without a condom, and selling blood but I have no self-control. In particular I cannot tolerate withdrawal symptoms and need a fix. If I have no money, blood selling is my only option. After taking the drugs, if I feel the need for sex then I go to brothel or mix with the 'station girls'. I have some problems in my penis, and maybe these can be transferred to them'.

He also mentioned that he used to go an NGO treatment centre but they refused to admit him because he has a blood shortage problem in his body. I found some addicts who confessed to sharing needles with their addict friends and in some cases with Indian drivers at Benapole. This was mainly when they ran out of needles from the hospital, dispensary or NGO office. Although some IVDUs share their needles due to lack of money, others regard it as a symbol of solidarity. Another, vulnerability is that sometimes they use a '*cocktail*' of drugs instead of heroin. In order to make cocktailing drugs, injecting drug users frequently mix up benzodiazepines (sedil, sedaxin) and anti-histamines (phenergan, avil) with the main drug (usually buprenorphine) to enhance and prolong the intoxicating effects, and alleviate the unpleasant side effects. Sometimes heroin smokers cannot get enough money together to buy heroin, which is expensive, and so they take a cocktail mixture of injectable liquid drugs which they believe to be more stimulating at a lower price.

Due to a good number of awareness and syringe distribution programmes by NGOs, many addicts are now using individual needles rather than sharing. But the problem is that they are taking the *cocktail* of drugs from a big ampoule with different needles and they cannot help but mix their blood into the drug cocktail and then pass it into another's body. So, although they are using individual syringes, they are vulnerable due to the sharing of the *cocktail* by everyone. One participant told me that there are some syringes found in the market which have been used but subsequently repackaged. According to him, people collect used syringes from waste bins in hospitals or clinics and sell them on the black market. They go to Dhaka to be re-packed and are then widely distributed. On the other hand, when people cannot afford a '*pata*' (a small wrap) of heroin, they share their money and jointly take the drugs. Sharing the same pipe for taking the '*mal*' (drugs) can cause the spread of infectious diseases such as hepatitis B or C. In addition, adding to fears that current and future heroin users in Bangladesh will convert from inhalers to injectors is the fact that most IVDUs in the neighbouring Indian state of Manipur and the neighbouring country Myanmar (Burma) were heroin addicts.

Transport Workers' Multi-Sexual Relationships It is also mentionable in many studies that truckers are an important client group for commercial sex workers as they spend extensive periods away from their families, which may contribute to them getting involved in new and different types of sexual relationships. These sexual behaviours placed them at high risk of STDs due to the long duration of infectiousness. There are also some Bangladeshi women living close to a truck stand who are not CSWs but are likely to be at higher risk than women in general due to their contacts with men at high risk for STDs, such as truckers and drug users. It was reported to me that many transport workers use a girl group-wise, so there is every chance of cross-infection from one body to another. As many girls have no scope to wash after sex, many of them suffer from skin infections, particularly in the genital area. These skin problems, and STDs generally, are then transmitted by the transport workers to their wives and other sex partners in their own village or home, where they also avoid using a condom. In other words, they provide a significant long-distance bridge for the spread of diseases such as HIV/AIDS. In Benapole, I found a few transport workers willing to discuss this. Jamal is one of them:

> 'I had sex with a street girl for the first time 3 years ago. Now we find many girls on the street and I have sex almost every night. I don't like condoms because they don't give the same pleasure. The girl doesn't insist on a condom because she needs the money. It doesn't matter to her how I am having sex with her! Although I have been facing some itching and other problems, when I see the girls I cannot control myself. We don't need to go to them, they come to us and ask about our needs. Many times we friends do sex with one girl. I've heard about the AIDS disease, that it can come to my body from the girls as I don't use condom. When I go to home after a while, I also have sex with local girls, such as my neighbour's bhabi [sister in law] who lives beside our house. We have had sex many times without condom in her house. She likes me and wants sex from me'.

Regarding protection, some transport workers claim that cannot buy condoms at night, and if they are offered one by the street girl, they feel embarrassed and insulted. Because they are paying, it seems to them that the girl should have no role in establishing 'the rules'. Moreover, most transport workers are non-literate and seem to have little idea about the proper utility of a condom and are concerned about personal satisfaction during intercourse. Mahbub at Benapole told me that *'it's a matter of pleasure; the "taste" is different with and without a condom. I don't like condoms; actually I cannot get a sense of what I am doing if I use a condom. Sometimes the girl asks me but I don't care. I want to the body touch but a condom acts like a veil'*.

Although many transport workers, particularly drivers, are involved in commercial sex, their wives do not know about their husbands' attitude. Some guess and forbid mixing with other women. But none of the drivers confess and the wives remain unaware of the risk they are running in having unprotected sex. In this regard, a truck driver, Idris at Khulna never says anything about his sexual exploits, in order to maintain marital stability and peace. He is confident that his wife is unaware of his extramarital relationships and before he returns home his sexual 'deficit' is filled.

4.3 HIV Knowledge and Consciousness

Knowledge of the ways in which HIV is transmitted is critical to adopting behaviours that prevent infection (UN, 2002). In the case of HIV infection prevention, many 'risk theories' have been developed by researchers to examine how people adopt behavioural change. In the public health discourse of HIV/AIDS, risk is usually considered as a certain type of lifestyle, such as unprotected sex as a risky behaviour (Campbell, 1997). From the late 1980s to the present, social science research has focused on surveys of risk-related sexual behaviour or the knowledge, attitudes, beliefs and practices (KABP) about sexuality which might be associated with the risk of HIV infection through quantifiable data (Cleland & Ferry, 1995). Different models have been formulated in this respect, such as KABP on condom use. But some authors (e.g. Denscombe, 1993) criticized these perspectives 'for treating individuals as free agents in terms of their response to risk and ignoring socio-cultural factors that constrain choice' (Teye, 2005; p. 66). Their argument is that individual risk perception is a social construct and depends not only on knowledge but also on controls that people feel they have on their own and with their partners. In the developing world, particularly in sub-Saharan Africa, gender and religious involvement play an important role in HIV/AIDS-related knowledge, attitudes and preventive behaviour (Lagarde et al., 2000; Takyi, 2003; Turmen, 2003).

The rapid sexual spread of the HIV virus worldwide has increased awareness of the importance of more effective control programmes for other STDs. Data on the general population show that there is widespread lack of knowledge and skills required to protect oneself and others: 30% of ever married women aged 15–49 years had never heard about HIV or AIDS, and only 11% have comprehensive knowledge about AIDS (NIPORT, 2016). Low awareness regarding HIV and AIDS among the general population is of concern as it may affect behaviours of potential clients of sex workers (NASP, 2016). In Bangladesh, there are many misconceptions about the transmission of STDs or HIV such as kissing, using contaminated clothes, sharing food and sharing toilets, or even being in a room with an HIV-infected person. Most had heard about syphilis and gonorrhoea, but they did have some misconceptions. Interestingly, when I asked sex workers anything about HIV, they responded quickly that 'it's a matter of needle-sharing'. On the other hand, when I asked drug users, they replied that prostitution is responsible. So, both of these vulnerable groups do not perceive themselves as 'risky' people for HIV dissemination. They prefer to blame others. This section will describe vulnerable people's knowledge and consciousness about STDs and HIV and also related reasons for their beliefs.

Sex Workers' HIV 'Knowledge' Although the infection rate of HIV/AIDS among the total population in Bangladesh is still very low, commercial sex workers' risk of being infected is considered to be significant. There is a high-risk HIV/AIDS transmission due to inadequate knowledge of the basic concepts of HIV/AIDS, low frequency of condom use and lack of treatment for STDs among sex workers. In the field, when I put HIV-related questions to my respondents, some became worried.

Sex workers particularly replied very promptly that they have no HIV in their body. Possibly they thought that, if they showed any curiosity about the disease, I would think that they were infected. Some replied in the same way about STDs thinking that I might identify them to the NGOs as 'bad' or 'sick' women. They also thought that, if they told me about their physical problems, word might get back to their customers through other girls. As a result, they always tried to keep the matter silent or ignore it. Some wanted to show themselves as 'fine' but demonized others due to local stigma.

When I asked the sex workers in different FGDs or in individual interviews about the ways in which HIV can be transmitted, many girls told me that they know everything about HIV transmission and means of prevention. Many were aware of the HIV epidemic in India. At Benapole, sex workers admit that AIDS is coming from neighbouring India and claim themselves to be disease free although their regular partners are Indian truckers. One girl (Popy) told me '*I have heard that about AIDS disease, but I've not met anybody who is infected. People say that AIDS has come from India. If I sleep with any of them it will happen. It can happen if I take shower in the same bathroom, or sleep together, or take food from the same plate*'. Some believe that HIV is a disease sent by Allah to punish 'sinful work' and that there are no effective medicines.

Although many girls have a fear of HIV, all are agreed that men do not appear to be concerned. When sex workers asked customers to use protection measures during intercourse, most defended themselves as either 'fresh' or 'virus free'. One hotel girl, Lima told me that '*when I ask them to use a condom use, some customers tell me that they have had a blood test and that they are "virus free"*'. On the other hand, if a girl tells the customer that she is sleeping with many men, and that she might be carrying the 'virus', many customers are prepared to take the risk and say, '*if I'm willing to catch your virus, what is your problem*?' As a result, some girls are recently feeling fearful about getting AIDS as their customers do not use proper protection. One of the Benapole girls, Sajeda told me in this regard that '*after I came to know about AIDS I am thinking that it may happen in my body because I never use a condom. I don't know whether it has appeared in my body or not*'. Some sex workers are so aware of the HIV issue that they refuse sex without a condom. I found one residence-based sex worker, Keya at Khulna who was very determined about HIV prevention:

> 'I love my son and I want to live for him. Maybe my life has no value to my customer. If I work with a customer without a condom, he will give me 100 taka at best, but he might give me the virus for my whole life. How many 100 taka I would need to spend for my treatment then? So, I don't want to be sick for not using a condom'.

Sometimes, girls face an inadequate supply of condom due to condom shortage in the market. They still needed to work with customers because of food costs and house rent. If condoms are unavailable or highly priced in the local market, then the health risk will increase. One girl in an FGD at Jessore told me that:

> 'If a condom is unavailable, then I have to either refuse the customer or have sex without condom because I have to survive. I will not give you a guarantee that I will not work if

condoms are unavailable. We will be forced to do that even if the disease appears, I would not care'.

Many outreach workers told me that condom use awareness among the girls has been increasing, but the lack of knowledge among new girls is causing them repeated STD problems. One peer educator, Khairun told me that many street girls do not assimilate HIV awareness programmes properly due to their illiteracy, or sometimes girls are irritated by hearing the message repeated. However, there are still misconceptions about condom use. Most commercial sex workers, though many have knowledge about HIV, cannot follow or practise the concept of 'safe sex'. Some girls think that out of five clients, if two are without a condom they will not contract the infection, or it will not be very strong. In practice, girls can persuade their older clients but the young ones refuse protection or offer extra money. Particularly among brothel-based and street-based sex workers, the level of knowledge is quite low due to lack of education, compared to the hotel- or residence-based sex workers. I noticed that, although awareness campaigns have been started on safe-sex practices in Bangladesh, there is still confusion about the effectiveness of that awareness among sex workers. It is very difficult to get the real picture from the girls about their condom use. Usually it depends on who is dominant, the sex worker or the customer. Here men's involvement should be incorporated more precisely in 'safer sex' practice.

Drug Users' Consciousness of HIV There are many different beliefs among drug users about the ways in which HIV is transmitted. This was obvious in one FGD with drug users in Jessore. Though some participants mentioned sexual contact without a condom, and needle sharing as risk factors, others emphasized different ideas:

> 'HIV can spread from urine, such as when people urinate in a same part of an open field. If others urinate there, they will be affected by HIV. AIDS is like chicken pox; spots will cover the whole body from head to toe. It's like a skin disease or leprosy. Some insects come out from the black spots'.

My field observation is that most drug users believe that HIV is caught only from sex workers. Their mindset is that HIV is a sex-related disease and that the reservoir of the virus is in the woman's body. They do not think primarily that needle sharing or blood selling can spread HIV. Two serious drug users, Badiul and Kabir, told me that '*AIDS comes from the 'magi' (prostitute). If I have sex with one, it will come, so I need to use a condom*'. Regarding needle sharing, many addicts confessed that they have shared needles in the past but now there are increasing levels of consciousness, and understanding of the dangers after NGO attempts to spread the word about HIV issues. However, as many drug addicts have unprotected sex, sexual diseases are common. I interviewed a drug addict, Rafiq in Khulna, who believes that he is suffering from AIDS because he has many STDs:

> 'I believe I am an AIDS patient as I have many sexual problems. I don't know how it comes through, maybe by girls or injection, but I am not taking treatment because I want to die to get relief from my sufferings'.

Lack of awareness of HIV/AIDS is considered an important risk factor. Levels of knowledge and awareness are very low among drug users, making them a vulnerable group for potential HIV infection. Many respondents have heard of the HIV virus but their knowledge of the symptoms and means of prevention are often inaccurate. It could be argued, therefore, that more effective mass media public health awareness campaigns about STDs and HIV/AIDS are needed.

Transport Workers' HIV Awareness Transport workers often ignore the danger of AIDS since they go to sex workers for pleasure or recreation. They appear to hold this aspect of their behaviour in a separate box from their understanding of risk. Here education or awareness-raising meets resistance because of a particular mentality and taste for excitement. Because AIDS patients are rare and mostly invisible in Bangladesh, many of those exhibiting risky behaviour do not feel any need to change. Truckers have their own beliefs about the origin and spread of HIV. One is that HIV is transmitted through mosquitoes and dogs and that people who have sex with their pets and then with a partner are spreading the virus. Another issue relates to physical appearance as demonstrated in the following extract:

> 'People say that those who have AIDS become slim, so I avoid slim girls. We can see who is suitable for us by their walking style. We suspect those who are very slim, so we take fat girls, not very fat but medium size. Otherwise we don't feel good'.

Many truckers told me that HIV/AIDS can come through sex with 'bad women'. Many transport workers think that brothels and street prostitutes have HIV, and that they are solely responsible for the spread of HIV in Bangladesh. One transport helper, Jasim, told me that as he does not use a condom, many germs including HIV have access to his body:

> 'Prostitutes have the AIDS disease germs because they mix with many people, so they carry the germs. The girls are mainly responsible as they are mixing with many people. Brothel and street girls have many customers, so they are at greatest risk'.

During an FGD with rickshaw pullers in Khulna, I was told that few use a condom and many do not believe in HIV. They think that all the messages are fake because their ancestors never used condoms. The girls use contraceptive pills, so they do not need to use a condom. Condoms might be important for brothel girls, because they mix with many customers and they might have sexual diseases, but not with other girls. When I asked a group of drivers at Benapole about their knowledge of sexually transmitted diseases, many participants said that itching, spots on the front side of the penis, a burning sensation during urination and gonorrhoea are considered as 'sex diseases'. According to them, condoms prevent the 'poison' going from the woman's body to the man's body. Some truckers also believe that if they use a condom with their sex partners, the girls will benefit, not the men. Some were embarrassed about discussing sexual problems or even to know about the HIV issue. They think that if they discuss it, they might be thought to be suffering from it themselves, with the inevitable loss of prestige. Some truckers said that they were not interested in hearing about HIV because they themselves had not faced this kind of disease. One of the drivers summed up his reserve: '*I will not speak about my secret things to a doctor, because he might think that I am a very bad person*'.

In this aspect, I talked with an outreach worker, Jabed who works with truckers in order to pass on HIV knowledge. He told me that many transport workers have physical symptoms of STIs but they have a tendency to hide the problem. Usually they say that one of their friends is suffering this type of problem and pretend to listen on his behalf. Although genital infections are common, they will only go to a doctor when they feel severe pain or experience serious problems. In an FGD with rickshaw pullers, mostly non-literate people, I found that they have little knowledge of condoms and AIDS. Like the truckers, many rickshaw pullers will not disclose their sexual problems to a qualified doctor, rather they visit back street healers or hawkers to solve their sexual diseases. I found one transport worker at Benapole, Mahbub who told me that risk-related feelings do not come into his mind before work. His mind is full of anticipation about sex and girls, but afterwards he repents because of not using a condom. He said that '*I cannot think that "sex without condom is wrong", though I believe the NGOs message, but before sex it doesn't come to my mind. I cannot remember that message because my brain doesn't work properly. Different feelings appear in my mind such as "how can I get the girl, will I get pleasure?" I forget about the future, but after sex I regret taking risks*'.

After contracting STDs, some truckers have come to see commercial sex as a 'sin' and have been thinking about making a commitment to refrain from it in future. Sukkur said that '*after contracting a penis disease, my mentality has become bad; I am always feeling gloomy and now I feel weak physically. I consider my disease as a punishment from God. So I am begging pardon from him*'. The present study's findings illustrate that the sexual behaviours, beliefs and stigma of a large proportion of the transport workers place them at high risk of STDs and HIV/AIDS. Communication strategies can increase clients' knowledge of AIDS and STD transmission and prevention is a critical first step. Initial efforts might employ communication materials including posters and brochures and outreach educational workers at commercial sex sites, as this strategy has proven useful elsewhere.

4.4 Health 'Risk Coping' Practices

Over the last 25 years, a variety of studies by social scientists have addressed questions about how people manage illness in their everyday lives. In the global South, evidence suggests that only a very small percentage of STD sufferers attend public facilities, the majority instead seeking clinical care from private doctors (Asthana, 1996), where both diagnosis and case management tend to be inadequate, or resort to self-medication. For instance, many sex workers in Madras believe that washing their genitals after intercourse in urine, soda water or lime juice will prevent STDs (Asthana & Oostvogels, 1996). Apart from the physical risk of complications, psychological responses to abortion are the most difficult to assess and evaluate (David & Lee, 2001). Legally, abortion is only permitted in Bangladesh to save the life of women, but menstrual regulation (MR) is legal, provided that pregnancy has not been confirmed (Streatfield, 2001). Due to social stigma and confidentiality issues, a large majority of induced abortions are by traditional methods and the result is often a long duration of continuous bleeding and a fever for more than 5 days (Bhuiya, Aziz, & Chowdhury, 2001). Chronic substance abuse has adverse and potentially life-threatening health consequences and detoxification is the first stage of treatment to protect the user from the withdrawal symptoms (Roehrich & Goldman, 1993). Persons detoxified from opioids, such as heroin, may be given a slower acting opioid, such as methadone, or clonidine, a nonopiate drug that decreases activity of the central nervous system (Lex, 2000). Some people have looked to alternative medicine, herbal remedies, homeopathy, acupuncture and hypnotherapy, and other therapies as alternatives to tranquillizers to help them withdraw (Hamlin & Hammersley, 1993). Other than treatment, addicts are required for the physical and psychological rehabilitation and care to attempt to change the environment to which the 'cured' addict returns. In this section, vulnerable people's varied health risk coping techniques will be described.

Sex Workers' Physical Risk Minimization Many commercial sex working girls take different techniques to avoid the physical risk in Bangladesh. In Bangladesh, as many sex workers do not have the opportunity to use a condom, some of them are involved with techniques to minimize their risk. Some use 'traditional' techniques in order to be risk free; some try to negotiate condom use according to their ability; some take homeopathic and herbal medicines; some take female contraceptives to prevent pregnancy; some have had an abortion and there are a few who wear an '*tabiz*' (amulet) to relieve menstrual pains. Although many STD patients get treatment, their partners do not have that opportunity and, as a result, they are infected. Some patients do not attach much importance to STDs and take only half a course of treatment. So, their cycle of STI infection continues. A list of the risk minimization techniques is described in the following.

'Traditional' Coping Strategies Many girls think that washing after sex is a means of reducing the risk of disease. In brothels, girls usually use water with soap and Savlon to clean themselves after sex. In the case of floating girls, they carry soap in a small bag and wash the vagina in the toilet. Sometimes they prefer to wash with warm water and apply Savlon after returning home to reduce the risk. Hotel sex workers follow 'traditional' techniques to be risk free. They try to urinate instantly after intercourse so that the semen comes out along with the urine. They think that if they sleep or lie on the bed after sex, pregnancy will develop. After urinating, they wash simply with water in most cases. Some girls use shampoo and soap after sex in an attempt to be risk free. In an FGD with hotel girls at Khulna, they said:

> 'When I feel I need to take this work without a condom, before sex I drink a lot of water, so that afterwards I can urinate. If the semen is left in the vagina, you have to urinate as soon as possible. Then all of the semen will come out'.

Many girls need to take customers without using a condom, so they are accustomed to using herbal or homeopathic medicines or '*kabiraji*' to protect themselves from sexual disease and unwanted pregnancy. Interestingly many are fearful about modern medical instruments for checking STDs. They are fearful of using contraceptive pills, which are said to affect their uterus. One hotel girl, Lime, told me at Khulna that she took homeopathic medicine to 'clear' a possible pregnancy; she became 'clean' and she started her menstruation again after one week. Moreover, some sex workers wear the *tabiz* to get relief from physical suffering. They have a strong superstitious belief that if they wear the *tabiz* their physical problems will disappear. I found one girl, Sapla at Baniashanta brothel, who was using a *tabiz* to relieve an abdominal pain. She bought it from a local *huzur* or religious man. Other girls in that brothel influenced her to collect it because many of them also using that kind of amulet. After using this *tabiz,* the problem has persisted but she believes that its effect will come after a few days. However, although many girls think that the condom is a symbol of women's security, they cannot force their customers to use it. Some girls try to check the customer's body and whether he has any scabies or any visible disease or not. Some sex workers check the penis of their clients for visible ulcers or discharge before engaging in sex, but in most cases the customers refuse. They also assume that their repeat or regular clients are free of AIDS and STDs, and believe that clients who look healthy and clean do not have any disease.

Female Contraceptive To reduce the risk of pregnancy, most girls use female contraceptives. Regarding female contraceptive use, the girls' fears mainly focus on the possibility of unwanted pregnancy rather than on STDs or HIV problems. To them, pregnancy is more important than disease because, if they become pregnant, they will immediately be identified as a sex worker by the community. It has been seen that many brothel or non-brothel-based sex workers are adapting to contraceptive methods most of the time. Apart from natural family planning like the calendar method, they use temporary methods such as pills, injections and implants, as well as permanent methods like bilateral tubectomy. Commercial sex working girls use contraceptives for different purposes along with condoms to keep themselves as 'safe' as possible. For the sake of their business, girls want to control their bleeding during menstruation periods by taking contraceptive pills and iron tablets, which control the hormone levels in the body, and also to avoid pregnancy. Some girls like pills, others prefer injections. One floating girl, Johura told me that '*I don't take the pill because it creates some side effects like vomiting. Instead I have injections which last for 3 months. They are safe, but with pills I cannot remember to take them properly*'. Many girls like the 'injection' method to escape blame and stigmatization. If they are seen to use the pill, people will say that 'you have no husband, so why you are on the pill?' But in the case of injections, nobody will know. A local healer in Benapole told me that he has many female STD patients who are using female contraceptives to prevent unwanted pregnancy. So, they have no pregnancy problems, but suffer from disease. When NGO doctors ask a girl about condom use, she will just reply 'yes' because girls do not usually trust the doctors, but they may subsequently confess to not using condoms consistently. They do not like to tell this fact to the doctor, so doctors need to behave in a friendly way to open a sex worker's window of recovery.

Abortion The majority of sex workers usually prefer to prevent unwanted pregnancies through contraceptive use rather than through abortion. Normally, as soon as a sex worker realizes that she is pregnant, she discusses the matter with her close friend or *babu* or *sardarni*. Then in most cases she contacts a local healer for an abortion. It is widely believed that local healers or midwives have more information about MR than an NGO doctor. Frequent performance of MR is common among commercial sex workers. Some sex workers have MRs performed two or three times per year, which causes excessive bleeding and weakness and leaves them in ill-health. Many floating girls are reluctant about condom use and forget to take their contraceptive pills on time. As a result, many of them need to have abortions for unwanted pregnancies. Almost every sex worker has had the experience of an abortion. One street girl, Chompa told me:

> 'Recently I was carrying a baby, I wanted to abort it, but who will give me the money to 'wash' it? I took some medicine to kill it but failed; right now I cannot 'work' properly because some customers know about my pregnancy. They make many bad comments. It's hampering my earnings'.

Sex working women have resorted to abortion to end unwanted pregnancies and often at considerable risk to personal health. However, social stigma attached to induced abortion may be a reason for not seeking safe abortion services by poor girls, since it may not remain confidential. If the local healers or clinic operators fail to abort the foetus, girls need to go to a large hospital and spend more money. But unfortunately none of the healers or clinic operators will take responsibility for any negative consequence or side effects caused by their treatment, the abortions or the MRs they perform.

Condom Negotiation Condoms are associated with contraception rather than disease prevention and are presumed to negatively affect sexual pleasure. As a result, they are rarely used in encounters between CSWs and their clients. Some call girls have a gentle approach about condom negotiation. They ask clients about virus transmission and its long-term impact. Recently, a few hotel or residence girls have been practising a technique to wear the condom by keeping it in the mouth in the name of oral sex. On the other hand, recently some brothel girls have been trying to develop a system informally that is 'payment first, sex after' for general customers to ensure condom use. After getting payment, the girl asks about condom use. If the customer does not agree, then they push him out of the brothel. However, some sex workers have the experience with female condoms but many do not like them because they are complicated to use, and many girls feel uneasy because semen remains and so the next customer can be infected with a previous one's STD.

Drug Users' Risk Minimizing Techniques and 'Coping' In Bangladesh, many drug addicts think that they will die from their addiction. This fear basically comes from their physical weakness and marginalized status. As a result, many addicts, mostly those who are involved with NGOs, are trying to engage with risk-reducing strategies like sexual abstinence, detoxification, the use of traditional medicine and treatments, drug rehabilitation training and even a preference for going to prison. This section will describe drug users' techniques for 'coping' with the health risk and their perceptions of obtaining relief from risk.

Treatment for Drug Addicts The traditional approach to detox is clinic based. Treatment is necessary for those abusers who have become most deeply involved with drugs. A drug user is considered as toxified for his or her physical and psychological dependence, which is exposed through withdrawal symptoms. In Bangladesh, low availability and limited success make the government-run detox clinics non-viable for most poor people. Many are used to being admitted to private drug treatment centres by their relatives. But most of these drug addiction treatment centres are run as for-profit businesses. Usually they have a casual or locum doctor (some centres have no staff doctor) and a number of unqualified staff, who give injections and act like a nurse. They are responsible for treatment, although standards are well below those that are desirable. When people see their sons or relatives becoming difficult and disturbed, they seek out this kind of treatment centre, which may also be considered as a 'gentle prison' for middle class people. Some addicts come here by their own choice and also leave similarly. In that case, the centre's role is merely to provide a commercial service like any other. Most expect that a patient will come again to their centre when the addiction crosses a certain threshold of suffering. When addicts arrive for the first time, they can bargain for their living expenses or room rent. Perhaps surprisingly, the centre staffs give patients a chance to take heroin for the 'last time'. There is an accusation that some private treatment centres even sell it at a market rate and a cynic might label some as heroin desire extending centres rather than centres for protection from heroin.

There are also a few NGO drop-in-centre (DIC) and treatment facilities where drug addicts are allowed to get treatment. Those that have been providing service for drug addicts have two kinds of treatment facilities: short and long term. The main goal of short-term treatment is to remove physical toxification and long-term treatment removes the psychological dependence on drugs. NGOs often give clonidine-type medicines to decrease withdrawal pains, along with some anti-histamine and sleeping tablets. During the follow-up period, they use a medicine named naltrazone to cut and block heroin toxicity. However, in the drug treatment centre, bath therapy is a commonly used technique for the drug patients in order to erase withdrawal symptoms. In a participant observation with drug patients in one NGO treatment centre, the centre manager told me that almost all drug patients have a mental demand for drugs for up to 72 h, then it decreases gradually. I met with some treatment taking drug patients in Khulna and one of them, Rahim said to me that they are learning some techniques to avoid the bad company which may 'pollute' them again.

Mind Control Lessons at Rehab Psychopharmacology often plays a key role by stabilizing patients so that gains can be made in psycho-social areas. A variety of therapies have been used in individual and group formats. Self-help groups, such as Narcotics Anonymous (NA), are very inexpensive and highly effective community resources that verbalize beliefs and values that encourage and facilitate responsible, insightful behaviour without the use of drugs. In Jessore, I found a few drug addict rehabilitation centres. In the rehabilitation centre, many mind control lessons have been provided for the drug patients to keep their mind stable, as it is thought that there is a close relationship between the mind and drugs, so by controlling emotions, it is possible to control the addiction. In addition, prayers are performed for relief from the world of addiction in future. They seek god's help for controlling their minds. The morning and afternoon prayers are:

> 'Almighty god, please bless addicts, please give us peace, so that we can change. Give us courage so that we can change whatever it is possible, and give us knowledge so that we can understand the difference!'

In the afternoon, there is a session in the rehab centre called 'face to face'. In this session, addicts openly criticize each other for wrong behaviour in the previous week, with reminders about the punishments for misdeeds in the programme. After their confession, punishments are meted out to increase their tolerance level as much as possible. Finally, they share their real life stories and miseries from drug addiction to increase their self-consciousness.

Prison Considered as an Opportunity for a 'New Life' Private detox treatment is expensive and prison is cheaper than detox. Sometimes the family sends the drug user to prison or the users get themselves into prison just to get off drugs. When some drug users feel that their life is spiralling out of control, they welcome a period in prison as a chance to get a 'new life' to come back into the mainstream. They think that if they can go to prison for a long time, their mind will be controlled against addiction and the body will become stable. As a result, many addicts see it as a form of safe custody. I found a few addicts who wanted to start a 'new life' or wanted to get relief from their health risks. One drug user, Lavlu who wants to get a new life by going to prison, said '*I want to come back again to normal life. I asked my wife to manage some money which will be needed for a 'new life'. I will go to jail and will stay there for minimum of 6 months. If I face bera (withdrawal symptoms) then I will have to tolerate it*'. In order to be admitted to prison, addicts first give money to the local police as a bribe to be brought into custody and falsely accused. Such addicts avoid the NGO treatment centre because it is only a short-term measure but in one of my participant observations in a rehab centre at Jessore, most of the participants did not agree with that concept. Their main logic is that if the addict could not control his mind, he will be addicted again. They think that in prison the addict will refrain from drugs for few days or months, but when they come out of jail they will go back to a heroin dealer.

Showing Fear or Gaining Awareness? During my field work, I was a participant observer in sessions in a drug rehabilitation centre in Jessore, where drug patients are being given treatment and training to know how to detach themselves from drugs when they leave the centre. I took the opportunity to organize some group discussions with participants. I asked which is more important for addicts to refrain from addiction: fear or gaining awareness? One man said that showing fear would work for him to leave addiction, but the rest disagreed with him. They thought that '*if you force someone leave addiction due to fear, he will be more crazy for that addiction. In all of our cases, family members tried several times to detain us and even beat us but we never left addiction. Instead the experience helped to fuel the addiction. If someone's consciousness is raised, it works promptly*'. In Khulna, some drug patients told me that they needed counselling on psychological issues to do with mind control. As they believe that drug addiction is a mental problem, so they thought that they should be given support with psychological problems. However, they also thought that if they have some form of vocational training, they would be more independent and this would help them to establish themselves on the outside. In addition, some drug users felt that there should be tablets to minimize the health risk from drug addiction.

Sexual Abstinence and Treatment Many recovering drug users have fallen back into the trap of drug taking. One claim by married addicts is that they are afraid that when they have recovered after treatment, they will not be able to 'perform' with their wife as previously. So their wife will be dissatisfied. In a participant observation in a private treatment centre at Jessore, I asked one of the treatment-taking patients (Rabiul) how he will manage and he told me:

> 'After getting treatment I will need to ask my wife, whom do you want? A man of sexual appetites or a drug-free man? If you wish a man with a sex drive, I will have to go to the pushers and take heroin again, then I will perform as before. If you want to a drug-free man, please give me six months to recover my sexual performance, which collapsed with my previous addiction. I believe my wife will wait for me'.

Many STD patients buy medicines from back street 'doctors' including homeopaths and herbal practitioners who claim that they can treat HIV. Pharmacists also recommend medicines for patients with STDs who cannot afford to visit a conventional doctor. These various healers understand the patients' mentality and have a very friendly attitude and a common touch. Patients are wary of recognized doctors, who may be suspicious of their character and try to explore their identity. I found one addict, Didar who is suffering many sexual problems. One of his relatives advised him to use soap and mustard oil on the penis for his disease which he is now doing. Moreover, I found one injecting drug user, Rassel at Khulna who used a 'hot water touch' to get back feeling in body parts which had become senseless due to frequent needle use. In order to prevent relapse, other than these addicts' coping techniques, local communities' attitudes towards the former addict need to be changed. As of now, addicts' returning to the community is likely to find themselves treated as ex-criminals. Negative public attitudes may drive former addicts back all the more quickly into the addict subgroup, with an even more confirmed anti-social attitude.

Transport Workers' Herbal Treatment Traditional healers are poor at recognizing sexual diseases and readily refer clients to modern health services. But not many poor people in Bangladesh seek treatment from licensed medical doctors. One worker, Zillur feels that his sex 'power' has been decreasing and now he cannot sustain intercourse for long. Another, Sukkur told me that his problems remain despite taking many medicines from local healers and some homeopathic and *kabiraji* (herbal) medicine. Unqualified medical practitioners like *kabiraj* (herbal practitioners) or homeopathic doctors usually start their businesses in a neighbourhood of transport terminals, brothels, slums and other areas where commercial sex workers congregate. As most of the vulnerable people like sex workers, drug users or transport workers are uneducated, they rarely inquire about the qualification of the doctors. Generally those who visit prostitutes and face symptoms such as loss of sexual drive, or physical changes in the penis, first consult an unregistered doctor for treatment and only seek the help of qualified practitioners in extreme cases. However, I did speak to transport workers and drug users who seek modern medical assistance from chemist shops and the 'doctors' (actually compounders or prescribing pharmacists) who work there. This use of medicine from local healers is common among participants in illicit sex and few go to medical practitioners due to the stigma. One local healer in Benapole told me that drivers and helpers are interested to take advice or medicine for short-term relief and then they visit doctors in Dhaka, who will prescribe antibiotics for STDs. Many people in Bangladesh, especially poor people and migrants, prefer traditional local healers because they have time to listen and to communicate in an understandable and sympathetic way.

4.5 Knowledge, Risk Minimization and Sustaining of HIV Risk

There is a significant difference in AIDS-related knowledge and behaviours among the population groups related to HIV prevention. In Bangladesh, there are certain social and cultural barriers to discussing and addressing HIV/AIDS, with the result that there is a low level of knowledge among the general public, including vulnerable people. As sex work and drug use are 'taboo' subjects, so people are reluctant openly to address the issues and are much more likely to hide their concerns about HIV. High rates of STDs and other health problems among vulnerable people reflect very low levels of condom usage by sexually active individuals in Bangladesh. Many sex workers and their clients particularly transport workers think that condom use is related to shyness and a symbol of sickness to some extent. Many male respondents feel stigmatized about condom access and usage or feel that there is no need to use a condom because their sexual partners are not sick.

This lack of acceptance of messages about safe sex among the male participants is due to their myths, beliefs and practices. I also found that many sex workers are not aware of the significance of STIs because some STDs are asymptomatic and, importantly, there is a fear and stigma attached to seeking care and treatment for a STD. Drug users do not have access to formal health care to counter their risk behaviour. Due to not having the financial ability to seek treatment and also having a negative community attitude, drug users cannot change their risky behaviour easily. There would seem to be an opportunity for the mass media to implement awareness programmes on the risks of needle sharing and its association with HIV. However, the consistent use of condoms is perhaps the single most important determining factor in controlling STIs among these vulnerable people. The growing incidence of HIV risk behaviour means that greater emphasis needs to be placed on methods that achieve disease prevention through proper health education and awareness raising. In the next chapter, I will examine the role of 'place' in channelling potential risk and the importance of geographical location in determining the hidden disaster of HIV in Bangladesh.

References

Alam, N., Rahman, M., Gausia, K., Yunus, M. D., Islam, N., Chaudhury, P., et al. (2007). Sexually transmitted infections and risk factors among truck stand workers in Dhaka, Bangladesh. *Sexually Transmitted Diseases, 34*(2), 99–103.

Asaduzzaman, M., Higuchi, M., Sarker, M. A., & Hamajima, N. (2016). Awareness and knowledge of HIV/AIDS among married women in rural Bangladesh and exposure to media: A secondary data analysis of the 2011 Bangladesh Demographic and Health Survey. *Nagoya Journal of Medical Science, 78*(1), 109–118.

Asthana, S. (1996). AIDS-related policies, legislation and programme implementation in India. *Health Policy and Planning, 11*(2), 184–197.

Asthana, S., & Oostvogels, R. (1996). Community participation in HIV prevention: Problems and prospects for community-based strategies among female sex workers in Madras. *Social Science and Medicine, 43*(2), 133–148.

Azim, T., Alam, M. S., Rahman, M., Sarker, M. S., Ahmed, G., Khan, M. R., et al. (2004). Impending concentrated HIV epidemic among injecting drug users in central Bangladesh (letter). *International Journal of STD & AIDS, 15*, 280–282.

Azim, T., Chowdhury, E. I., Reza, M., Ahmed, M., Uddin, M. T., Khan, R., et al. (2006). Vulnerability to HIV infection among sex worker and non-sex worker female injecting drug users in Dhaka, Bangladesh: Evidence from the baseline survey of a cohort study. *Harm Reduction Journal, 3*(33).

Bagley, C., Kadri, S., Shahnaz, A., Simkhada, P., & King, K. (2017). Commercialised sexual exploitation of children, adolescents and women: Health and social structure in Bangladesh. *Advance in Applied Sociology, 7*(4), 137–150.

Bhuiya, A., Aziz, A., & Chowdhury, M. (2001). Ordeal of women for induced abortion in a rural area of Bangladesh. *Journal of Health, Population and Nutrition, 19*(4), 281–290.

Brown, M. (1995). Ironies of distance: An ongoing critique of the geographies of AIDS. *Environment and Planning D: Society and Space, 13*(2), 159–183.

Bwayo, J., Plummer, F., Omari, M., Mutere, A., Moses, S., Ndinya-Achola, J., et al. (1994). Human immunodeficiency virus infection in long distance truck drivers in East Africa. *Archives of Internal Medicine, 154*, 1391–1396.

Caldwell, B., Pieris, I., Barkat-e-Khuda, C., Caldwell, J., & Caldwell, P. (1999). Sexual regimes and sexual networking: The risk of an HIV/AIDS epidemic in Bangladesh. *Social Science and Medicine, 48*(8), 1103–1116.

Campbell, C. (1997). Migrancy, masculine identities and AIDS: The psychosocial context of HIV transmission on the South African gold mines. *Social Science and Medicine, 45*(2), 273–281.

Cheng, S. (2005). Popularising purity: Gender, sexuality and nationalism in HIV/AIDS prevention for South Korean youths. *Asia Pacific Viewpoint, 46*(1), 7–20.

Choudhary, M. S., & Rahman, M. (2014). Knowledge, awareness and perception about HIV/AIDS among primary school teachers in Bangladesh. *Bangladesh Journal of Medical Science, 13*(2), 145–149.

Choudhury, M. R., Rahman, A. S. M., & Moniruzzaman, M. D. (1989). Serological investigations for a detection of HIV infection in Bangladesh. *Bangladesh Armed Forces Medical Journal, 13*(1), 1–9.

Cianelli, R., Villegas, N., Lawson, S., Ferrer, L., Kaelber, L., Peragallo, N., et al. (2013). Unique factors that place older Hispanic women at risk for HIV: Intimate partner violence, machismo, and marianismo. *Journal of the Association of Nurses in AIDS Care, 24*(4), 341–354.

Cleland, J., & Ferry, B. (Eds.). (1995). *Sexual behavior and AIDS in the developing world*. London, UK: Taylor & Francis.

Cliff, A. D., & Haggett, P. (1988). *Atlas of disease distributions: Analytic approaches to epidemiological data*. New York, NY: Blackwell.

Craddock, S. (2000). Disease, social identity, and risk: Rethinking the geography of AIDS. *Transaction of the Institute of British geographers, 25*(2), 153–168.

David, H. P., & Lee, E. (2001). Abortion and its health effects. In J. Worell (Ed.), *Encyclopedia of women and gender* (Vol. 1). New York, NY: Academic Press.

Denscombe, M. (1993). Personal health and the social psychology of risk taking. *Health Education Research, 8*, 505–517.

Ellis, M., & Muschkin, C. (1996). Migration of persons with AIDS-a search for support from elderly parents? *Social Science and Medicine, 43*(7), 1109–1118.

FHI. (2006). *Summary of an assessment of sexual behaviour of men in Bangladesh: A methodological experiment*. Dhaka, Bangladesh: Family Health International (FHI).

Frewer, L. J. (1999). Public risk perceptions and risk communication. In P. Bennett & K. Calman (Eds.), *Risk communication and public health*. New York, NY: Oxford University Press.

Gagnon, A. J., Merry, L., Bocking, J., Rosenberg, E., & Oxman-Martinez, J. (2010). South Asian migrant women and HIV/STIs: Knowledge, attitudes and practices and the role of sexual power. *Health and Place, 16*(1), 10–15.

Gangopadhyay, D. N., Chanda, M., Sarkar, K., Niyogi, S. K., et al. (2005). Evaluation of sexually transmitted diseases/human immunodeficiency virus intervention programs for sex workers in Calcutta, India. *Sexually Transmitted Diseases, 32*(11), 680–684.

Gazi, R., Khan, S. I., Haseen, F., Sarma, H., Islam, M. A., Wirtz, A. L., et al. (2009). Young clients of hotel-based sex workers in Bangladesh: Vulnerability to HIV, risk perceptions, and expressed needs for interventions. *International Journal of Sexual Health, 21*(3), 167–182.

George, S., Jacob, M., John, T. J., Jain, M. K., Nathan, N., Rao, P. S., et al. (1997). A case-control analysis of risk factors in HIV transmission in South India. *Journal of Acquired Immune Deficiency Syndrome and Human Retrovirology, 14*, 290–293.

Gibney, L., Macaluso, M., Kirk, K., Hassan, M. S., Schwebke, J., Vermund, S. H., et al. (2001). Prevalence of infectious diseases in Bangladeshi women living adjacent to a truck stand: HIV/STD/hepatitis/genital tract infections. *Sexually Transmitted Infections, 77*, 344–350.

Gibney, L., Saquib, N., Macaluso, M., Hasan, K. N., Aziz, M. M., Khan, A. Y. M. H., et al. (2002). STD in Bangladesh's trucking industry: Prevalence and risk factors. *Sexually Transmitted Infections, 78*, 31–36.

Gibney, L., Saquib, N., & Metzger, J. (2003). Behavioural risk factors for STD/HIV transmission in Bangladesh's trucking industry. *Social Science and Medicine, 56*(7), 1411–1424.

Goldenberg, S. M., Strathdee, S. A., Gallardo, M., Nguyen, L., Lozada, R., Semple, S. J., et al. (2011). How important are venue-based HIV risks among male clients of female sex workers? A mixed methods analysis of the risk environment in nightlife venues in Tijuana, Mexico. *Health and Place, 17*(3), 748–756.

Gould, P. (1993). *The slow plague: A geography of the AIDS pandemic*. Oxford, UK: Blackwell.

Hamlin, M., & Hammersley, D. (1993). Managing benzodiazepine withdrawal. In G. Bennett (Ed.), *Treating drug abusers* (pp. 91–114). London, UK: Routledge.

Haque, M. A., Hossain, M., Chowdhury, M., & Uddin, M. J. (2018). Factors associated with knowledge and awareness of HIV/AIDS among married women in Bangladesh: Evidence from a nationally representative survey. *Journal of Social Aspects of HIV/AIDS Research Alliance, 15*(1), 121–127.

Hawkes, S. (1992). Travel and HIV/AIDS. *AIDS care, 4*(4), 446–449.

Hossain, M., Mani, K. K., Sidik, S. M., Shahar, H. K., & Islam, R. (2014). Knowledge and awareness about STDs among women in Bangladesh. *BMC Public Health, 14*, 775.

Islam, A. K. M. S. (2006). *The integration of Islamic values into drug demand reduction strategies-prevention, In the publication of Department of Narcotics Control, on the International day against drug abuse and illicit trafficking*. Dhaka, Bangladesh: Government of Bangladesh.

Islam, M. M., & Conigrave, K. M. (2008). HIV and sexual risk behaviors among recognized high-risk groups in Bangladesh: Need for a comprehensive prevention program. *International Journal of Infectious Diseases, 12*(4), 363–370.

Islam, S. M. S., Biswas, T., Bhuiyan, F. A., Islam, M. S., Rahman, M. M., & Nessa, H. (2015). Injecting drug users and their health seeking behaviour: A cross-sectional study in Dhaka, Bangladesh. *Journal of Addiction, 2015*, 8 pages.

Jenkins, C., & Rahman, H. (2002). Rapidly changing conditions in the brothels of Bangladesh: Impact on HIV/STD. *AIDS Education and Prevention, 14*(Suppl. A), 97–106.

Kabir, M. A., Goh, K. L., Kamal, S. M. M., Khan, M. M. H., & Kazembe, L. (2013). Tobacco smoking and its association with illicit drug use among young men aged 15–24 years living in urban slums of Bangladesh. *PLoS ONE, 8*, 7. (e68728).

Kalipeni, E., Craddock, S., Oppong, J. R., & Ghosh, J. (Eds.). (2004). *HIV and AIDS in Africa: Beyond epidemiology*. Oxford, UK: Blackwell.

Kearns, R. A. (1996). AIDS and medical geography: Embracing the other? *Progress in Human Geography, 20*(1), 123–131.

Khanam, R., Reza, M., Ahmed, D., Rahman, M., Alam, M. S., Sultana, S., et al. (2017). Sexually transmitted infections and associated risk factors among street-based and residence-based female sex workers in Dhaka, Bangladesh. *Sexually Transmitted Diseases, 44*(1), 22–29.

Khosla, N. (2009). HIV/AIDS interventions in Bangladesh: What can application of a social exclusion framework tell us? *Journal of Health, Population and Nutrition, 27*(4), 587–597.

Kuhanen, J. (2010). Sexualised space, sexual networking & the emergence of AIDS in Rakai, Uganda. *Health and Place, 16*(2), 226–235.

Lacerda, R., Gravato, N., McFarland, W., Rutherford, G., Iskrant, K., Stall, R., et al. (1997). Truck drivers in Brazil: Prevalence of HIV and other sexually transmitted diseases, risk behavior and potential for spread of infection. *AIDS, 11*(Suppl. 1), S15–S19.

Lagarde, E. M., Enel, C., Seck, K., Gueye-Ndiaye, A., Piau, J-P., Pison, G., et al. (2000). Religion and protective behaviours towards AIDS in rural Senegal. *AIDS, 14*(13), 2027–2033.

Larance, B., Ambekar, A., Azim, T., Murthy, P., Panda, S., Degenhardt, L., et al. (2011). The availability, diversion and injection of pharmaceutical opioids in South Asia. *Drug and Alcohol Review, 30*, 246–254.

Lawrence, J., Kearns, R. A., Park, J., Bryder, L., & Worth, H. (2008). Discourses of disease: Representations of tuberculosis within New Zealand newspapers 2002–2004. *Social Science and Medicine, 66*(3), 727–739.

Learmonth, A. T. A. (1952). The medical geography of India: An approach to the problem. In K. Kuriyan (Ed.), *The Indian geographical society, The silver jubilee volume* (pp. 201–202). Madras, India: Indian Geographical Society.

Lex, B. W. (2000). Gender and cultural influences on substance abuse. In R. M. Eisler & M. Hersen (Eds.), *Handbook of gender, culture and health* (pp. 255–297). Hillsdale, NJ: Lawrence Publishers.

Lindquist, J. (2005). Organizing AIDS in the borderless world: A case study from the Indonesia-Malaysia-Singapore growth triangle. *Asia Pacific Viewpoint, 46*(1), 49–63.

Luginaah, I. (2008). Local gin (akpeteshie) and HIV/AIDS in the Upper West Region of Ghana: The need for preventive health policy. *Health and Place, 14*(4), 806–816.

Madise, N. J., Ziraba, A. K., Inungu, J., Khamadi, S. A., Ezeh, A., Zulu, E. M., et al. (2012). Are slum dwellers at heightened risk of HIV infection than other urban residents? Evidence from population-based HIV prevalence surveys in Kenya. *Health and Place, 18*(5), 1144–1152.

Marshall, B. D. L., Kerr, T., Shoveller, J. A., Patterson, T. L., Buxton, J. A., & Wood, E. (2009). Homelessness and unstable housing associated with an increased risk of HIV and STI transmission among street-involved youth. *Health and Place, 15*(3), 783–790.

Marten, L. (2005). Commercial sex workers: Victims, vectors or fighters of the HIV epidemic in Cambodia? *Asia Pacific Viewpoint, 46*(1), 21–34.

May, J. M. (1958). *The ecology of human disease*. New York, NY: MD Publication.

Mbugua, G. G., Muthami, L. N., Mutura, C. W., Oogo, S. A., Waiyaki, P. G., Linden, C. P., et al. (1995). Epidemiology of HIV infection among long distance truck drivers in Kenya. *East African Medical Journal, 72*(8), 515–518.

Mondal, N. I., Islam, R., Rahman, O., Rahman, S., & Hoque, N. (2012). Determinants of HIV/AIDS awareness among garments workers in Dhaka city, Bangladesh. *World Journal of AIDS, 2*(4), 312.

Mondal, N. I., Takaku, H., Ohkusa, Y., Sugawara, T., & Okabe, N. (2009). HIV/AIDS acquisition and transmission in Bangladesh: Turning to the concentrated epidemic. *Japanese Journal of Infectious Disease, 62*(2), 111–119.

Moran, D. (2005). The Geography of HIV/AIDS in Russia: Risk and vulnerability in transition. *Eurasian Geography and Economics, 46*(7), 525–551.

Morris, K. A., & Aftab, A. (2012). Tobacco and substance use: Perceptions and practices among men in Bagnibari, Bangladesh. *BRAC University Journal*, (Special Issue), 61–70.

Murray, A., & Robinson, T. (1996). Minding your peers and queers: Female sex workers in the AIDS discourse in Australia and South-east Asia, Gender. *Place and Culture, 3*(1), 43–59.

NASP. (2016). *Fourth National Strategic Plan for HIV and AIDS response, National AIDS/STD Program*. Dhaka, Bangladesh: Ministry of Health and Family Welfare, Government of Bangladesh.

Nessa, K., Waris, S. A., Alam, A., Huq, M., Nahar, S., et al. (2005). Sexually transmitted infections among brothel-based sex workers in Bangladesh: High prevalence of asymptomatic infection. *Sexually Transmitted Diseases, 32*(1), 13–19.

Nessa, K., Waris, S. A., Sultan, Z., Monira, S., et al. (2004). Epidemiology and etiology of sexually transmitted infection among hotel-based sex workers in Dhaka, Bangladesh. *Journal of Clinical Microbiology, 42*(2), 618–621.

NIPORT. (2016). *Bangladesh demographic and health survey 2014*. Dhaka, Bangladesh and Rockville, MD: NIPORT, Mitra and Associates, and ICF International.

Parkin, S., & Coomber, R. (2011). Public injecting drug use and the social production of harmful practice in high-rise tower blocks (London, UK): A Lefebvrian analysis. *Health and Place, 17*(3), 717–726.

Patton, C. (1994). *Last Served?* Taylor & Francis, London, UK: Gendering the HIV pandemic.

Pyle, G. F. (1969). The diffusion of cholera in the United States in the nineteenth century. *Geographical Analysis, 1*, 59–75.

Rahman, M., Alam, A., Nessa, K., Hossain, A., et al. (2000). Etiology of sexually transmitted infections among street-based female sex workers in Dhaka, Bangladesh. *Journal of Clinical Microbiology, 38*(3), 1244–1246.

Rhodes, T., & Quirk, A. (1996). Heroin, risk and sexual safety. In T. Rhodes & R. Hartnoll (Eds.), *AIDS, drugs and prevention: Perspectives on individual and community action*. London, UK: Routledge.

Roehrich, L., & Goldman, M. S. (1993). Experience-dependent neuropsychological recovery and the treatment of alcoholism. *Journal of Consulting and Clinical Psychology, 61*(5), 812–821.

Sabin, K. M. (1998). *A study of sexually transmitted infections and associated factors in slum communities of Dhaka, Bangladesh*. PhD dissertation, The school of Hygiene and Public Health. Baltimore, MD: The Johns Hopkins University.

Sarkar, K., Bal, B., Mukherjee, R., Niyogi, S. K., Saha, M. K., & Bhattacharya, S. K. (2005). Epidemiology of HIV infection among brothel-based sex workers in Kolkata, India. *Journal of Health, Population and Nutrition, 23*(3), 231–235.

Sarkar, S., Islam, N., Durandin, F., et al. (1998). Low HIV and high STD among commercial sex workers in a brothel in Bangladesh: Scope for prevention of larger epidemic. *International Journal of STD & AIDS, 9*, 45–47.

Shaw, M., Dorling, D., & Mitchell, R. (2002). *Health, place and society*. Essex, UK: Pearson.

Sheikh, M. T., Uddin, M. N., & Khan, J. R. (2017). A comprehensive analysis of trends and determinants of HIV/AIDS knowledge among the Bangladeshi women based on Bangladesh Demographic and Health Surveys, 2007–2014. *Archives of Public Health, 75*(1), 59.

Singh, Y. N., & Malaviya, A. N. (1994). Long distance trucks drivers in India: HIV infection and their possible role in disseminating HIV into rural areas. *International Journal of STD & AIDS, 5*(2), 137–138.

Stamp, L. D. (1964). *The geography of life and death, Ithaca*. New York, UK: Cornell University Press.

Streatfield, P. K. (2001). Role of abortion in fertility control, editorial. *Journal of Health, Population and Nutrition, 19*(4), 265–267.

Takahashi, L. M., & Dear, M. J. (1997). The changing dynamics of community opposition to human service facilities. *Journal of the American Planning Association, 63*(1), 79–93.

Takyi, B. K. (2003). Religion and women's health in Ghana: Insights into HIV/AIDS preventive and protective behavior. *Social Science and Medicine, 56*(6), 1221–1235.

Tempalski, B., & McQuie, H. (2009). Drugscapes and the role of place and space in injection drug use-related HIV risk environments. *International Journal of Drug Policy, 20*(1), 4–13.

Teye, J. K. (2005). Condom use as a means of HIV/AIDS prevention and fertility control among the Krobos of Ghana. *Norwegian Journal of Geography, 59*, 65–73.

Thomas, R. W. (1992). *Geo medical systems: Intervention and control*. London, UK: Routledge.

Turmen, T. (2003). Gender and HIV/AIDS. *International Journal of Gynecology and Obstetrics, 82*(3), 411–418.

UN. (2002). *HIV/AIDS awareness and behaviour, United Nations (UN), Department of Economic and Social affairs, Population division*. New York, NY: United Nations.

Werb, D., Kerr, T., Fast, D., Qi, J., Montaner, J. S. G., & Wood, E. (2010). Drug-related risks among street youth in two neighborhoods in a Canadian setting. *Health and Place, 16*(5), 1061–1067.

Yaya, S., Bishwajit, G., Danhoundo, G., Shah, V., & Ekholuenetale, M. (2016). Trends and determinants of HIV/AIDS knowledge among women in Bangladesh. *BMC Public Health, 16*(1), 812.

Chapter 5
Stigmatized Place, Mobility and HIV Risk Channelling

5.1 Introduction

Sense of place is one of the ways in which geographers have attempted to approach the complex linkages between people's well-being and location (McCreanor et al., 2006). Place is not only physical, but is deeply connected to the self (Wiersma, 2008; Williams, 2002) and a lived phenomenon (Lorimer, 2019). Moreover, geographers have long distinguished between space and place, emphasizing that place is more than a physical location or container in which events unfold (Wiles et al., 2009). Rather, place could be thought of as a dynamic process invested with integrated physical, social, emotional and symbolic aspects which interact at a range of different scales (Massey, 1999; Wiles, 2005). There is a long history of interest in place and health in the geography of health (Dunn & Cummins, 2007). The concept of place, and links between place and health, is pressing the developments in contemporary health philosophy. One such development is the emergence of a conceptual model firmly based on ideas of health rather than medicine (Kearns, 1997) in which a socio-ecological model of health (argument of White, 1981) involves an interactive set of relationships between a population and their social, cultural and physical environment. Mohan (2000) thinks that the geography of health has emphasized the emergence of the theme of place for the sensitivity of difference. Curtis (2004) thinks that the geography of health is focused on the ways that the health of populations is differentiated between places and the range of factors that explain these differences.

Based on a thematic evolution, Kearns and Moon (2002) argue that there are now three themes that characterize contemporary health geography. These can be summarized as social constructions of place, the utility and greater awareness of sociocultural theory and the evolution of a critical geography of health (see also Parr, 2004). Relational natures of people's place experiences are the focus of much recent health geography research (Kaley, Hatton, & Milligan, 2019). However, the notion

A. Paul, *HIV/AIDS in Bangladesh*, Global Perspectives on Health Geography,
https://doi.org/10.1007/978-3-030-57650-9_5

of therapeutic landscapes has been embraced by health geographers and yielded a significant, and growing, body of recent research (Kearns & Milligan, 2020). Some studies in health geography (Dyck & Dossa, 2007; Gesler, 2005; Smyth, 2005) mention that the physical, social and symbolic landscapes of therapeutic environments including places, spaces and networks serve to regulate and normalize certain kinds of behaviour and the role of everyday activity in producing meanings and experiences of space as 'healthy'. Health geographers have been most active in the analysis of smaller unit areas (Andrews, Cutchin, McCracken, Phillips, & Wiles, 2007) and they have turned to narratives as a way to engage with the everyday, situated experiences of people in place (e.g. Kearns, 1997). They are interested in studying health in a particular place or in making comparisons between places and studying health events in a set of places (Gatrell, 2002). Importantly, geographers point out that discourses of health and illness are particularly powerful agents in the construction of places (for example, Craddock, 2000; Gesler, 1998; Kearns, 1998; Kearns & Gesler, 1998) as well as individual bodies (Brown, 1995; Dorn & Laws, 1994; Parr, 2002; Philo, 2000). Some works (Del Casino Jr., 2004; Dyck & Dossa, 2007; Elliott & Gillie, 1998) have focused on the embeddedness of their health practices in place, bringing to bear a conception of place that dismisses a notion of boundedness. Andrews and Evans (2008) and Cutchin (2007) illustrate a way to collect information about place and landscape and then interpret how the processes that create and re-create them bring about the situation that an epidemiologist typically investigates.

The majority of existing epidemiological research on place and health has focused on a single spatial scale, generally that of local areas or 'neighbourhoods' (Cummins, Curtis, Diez-Roux, & Macintyre, 2007). Few studies (Feldacker, Emch, & Ennett, 2010; Kuhanen, 2010; Rhodes, 2009; Tempalski & McQuie, 2009) discussed geographic place-based understanding identify the risk environments comprising interactions between individuals and environments in the context of HIV. A number of studies have been made to understand the impetus and reasons for migration (Saggurti, Mahapatra, Swain, & Jain, 2011; Vearey, Palmary, Thomas, Nunez, & Drimie, 2010; Wood et al., 2000) in spreading HIV from high-risk populations to low prevalence areas. In the context of Africa, some geographic research of HIV (Aase & Agyei-Mensah, 2005; Ferguson & Morris, 2007; Kesby, 2000; Mayer, 2005; Oppong, 1998) highlighted the roles of 'vulnerable places' in HIV transmission and the attitudes of vulnerable groups. In the last decade, some research (Goldenberg et al., 2011; Ivsins, Benoit, Kobayashi, & Boyd, 2019; Jennings, Woods, & Curriero, 2013) investigated the issues of venues, livelihoods, environmental dimensions, marginalization, risk and harm in HIV discourse.

There has been some critical reflection among geographers about a focus on mapping disease incidence, or showing the migration flows, without recognizing the social dimensions of the illness (Brown, 1995; Craddock, 2000). Following Wilton (1996), space is important in terms of material conditions for people living with HIV/AIDS or vulnerable to this health risk in the development of their daily paths and their shifting social networks, where people's daily worlds and routines changed after diagnosis. Bangladesh has been recognized as one of the countries in

Asia where HIV/AIDS infections are increasing rapidly (Hossain, 2007). It has been discussed in the previous chapters that the AIDS pandemic could be set to explode in Bangladesh in the near future because of the marginalized 'lifeworlds' of vulnerable people and their stigmatized identity, along with their high-risk behaviour and low awareness. In addition, the high mobility of vulnerable people particularly in border areas has a role in the potential channelling of health risk and it is another major threat in the context of HIV/AIDS in Bangladesh (Paul, Atkins, & Dunn, 2012). In other words, the notions of 'place' and 'borders' play an important role in shaping responses to the challenges of HIV/AIDS. In this chapter, I will discuss how place is playing a role in health risk, particularly the HIV threat among vulnerable communities (pictures shown in Photo Series 5.1 as symbolic). Secondly, the risk channelling role of border towns and land ports will be described in terms of their implications for women. Finally, the chapter will examine the hidden HIV cases in Bangladesh and will draw on relevant case studies.

Photo Series 5.1 Risk place, mobility and Indo-Bangladesh border

5.2 Place as a 'Risky' and 'Safe'

By tradition, geographers have been interested in the character of places. The importance of place, more recently, has been reasserted through the development of the sense of place concept (Tuan, 1991). Eyles (1985) has extended this idea, pointing to the relationship between an individual's place-in-the-world and experience of place. Shooting galleries contain all of the risk factors for the drug addicts which together make a 'risk environment'. Following Rhodes, Singer, Bourgois, Friedman, and Strathdee (2005), 'risk environment' is the place where a variety of factors combine to increase the possibility of HIV transmission. Epidemiological research identifies shooting galleries as physical environments in which injectors gather to inject drugs, and associates such places with an elevated risk of HIV transmission, particularly in galleries in which injecting equipment is rented or stored for re-use (Carlson, 2000; Des Jarlais & Friedman, 1990). In other words, injecting in public or semi-public places has in turn been associated with urban disadvantage, homelessness and a fear of police arrest resulting from high-profile policing practices (Bourgois, 1998; Maher & Dixon, 1999). In recent discourse about drug users' 'place', some studies (Ivsins et al., 2019; Mass et al., 2007; Parkin & Coomber, 2011; Rhew, Hawkins, & Oesterle, 2011; Werb et al., 2010) discussed that geographic place, poor urban neighbourhood become the environmental risk factor for HIV transmission among IDUs. They mentioned the negative effect of place on health risks which are implicated in experiences of structural (and physical) violence and marginalization. A Save the Children (2012) report shows that drug users in Bangladesh also reported facing high levels of violence. The notion of risk is profoundly gendered (Wilton, 1994). To explore the narratives of women who have experienced violence in different places while working as prostitutes, O'Neill (1996) mentioned that some women experience physical assaults.

In recent years, there has been a rapid growth of prostitutes in the cities and towns of Bangladesh particularly in hotels. A study found that about 39% of street-based sex workers faced sexual violence (Save the Children, 2012). In recent concept on sex workers 'place', a few studies (Gutierrez-Garza, 2013; Marshall, 2017; Marshall et al., 2009; McNeil, Shannon, Shaver, Kerr, & Small, 2014; Phipps, Ringrose, Renold, & Jackson, 2018; Walker, 2017) explored the relationship between place and violence upon sex worker's everyday life which play role in transmission of HIV and STI risk through sexual harassment. They illustrated how sex workers face multiple vulnerabilities, including discrimination, criminalization and different types of sexual violence every day from customers, police and the public due to the place factor. Moreover, study on long-distance truck drivers has explored their sexual cultures by reviewing the African (i.e. Nigeria, Zimbabwe and Kenya) and Asian (i.e. India, Thailand) literature on truckers (Jackson, Rakwar, Richardson, et al., 1997; Morris, Podhisita, Wawer, & Handcock, 1996; Orubuloye, Caldwell, & Caldwell, 1993; Rao, Pilli, Rao, & Chalam, 1999). As part of the global but uneven spread of the HIV/AIDS epidemic, notions of 'place' for drug users, sex workers and transport workers play an important role in shaping responses to the challenges of HIV/AIDS. This section will look at drug addicts' 'own place', sex workers' perceptions of safe or non-safe places and transport workers' preferred sites, which are key to the HIV threat facing them.

The Stigma of Place of Drug Users The type of drug use by an addict depends on their 'own place' to take drugs safely. After purchasing heroin or some other injectable substance in a local drug selling area, addicts need to leave the place quickly to avoid the harassment by police and they look for 'safe' place for taking drugs. Drug dealings usually take place on the street or some public abandoned establishment. In Bangladesh, drug addicts have their 'own place' where they can take drugs freely and without fear. On the other hand, there are bad places where they face harassment. In their own safe haven they not only take drugs but also exchange feelings. This variety of places forms one dimension of their 'map' of health hazards. When forced into the melée of bad places they are more likely to be at risk, for instance, from needle sharing or staying in unhygienic conditions, but, unfortunately, there are very few places that addicts can call their own. A concomitant fear of the carriage of used needles and syringes as constituting evidence of possession or providing a rationale for increased police interest also discourages injecting drug users from using local pharmacies for the purchase of clean injecting equipment. If they sit in a public place, they will soon attract threats, and because the police try to arrest addicts at the point of sale, most prefer to take their drugs in quieter places where they will not be disturbed. This may be a garden, slum area, graveyard or an abandoned building. One injecting drug user at Khulna, Salauddin, feels that addicts need to take drugs in a hidden place, as secret as possible. Sometimes they take drugs with friends in the market or in an open field in secret and they need to be careful not to be noticed.

Regarding heroin or other drug selling points, drug users told me that there are many such places in both Jessore and Khulna. Sometimes local people resist the dealers but after a few days they come back and sell again. In both towns, drugs are sold close to educational institutions, the railway station, market, bus terminal and in nearby villages. The availability of drugs in a society is an important and indirect indication of the heroin use pattern in a country. If heroin becomes more widely available and cheap, it is clear that there will be many more users. The easy availability of drugs is a prime cause of the growing number of drug abusers in Bangladesh. During my participant observation in a commercial drug treatment centre at Jessore, I met with one field level worker who collects patients for his centre and has been working with addicts for a long time. He thinks that drug addiction is becoming more of a problem than HIV because so many youths take drugs at one time or another and they are being 'polluted', which can lead to death. He said:

> 'In the Jessore region you will see huge numbers of people are involved with drugs, particularly heroin. Most of our heroin is coming from India. You can easily buy it, from a tea stall even. I have found in many places even rural settlements are affected by drugs'.

Regarding mobility and the need for drugs, many users admitted to me that the level of desperation is at times so great that addicts do not care about place or threat to life. When they face police harassment in Jessore then they collect their drugs from Benapole, which means more business for the dealers there. In the south-western region of Bangladesh, Benapole is considered to be one of the most 'trust-worthy' heroin collection points. Drug users from different parts of Bangladesh know the reputation of Benapole for its 'reliability'. If necessary, they also cross the rather porous border with India to buy the product. In Jessore, one FGD participant told me:

> 'When we feel a crisis, we go to Benapole, because it is a 'base' station for heroin, we take the utmost risks to get the product on time. We don't care about the river, jungle or even the border security force. When the BDR or BSF notice us, they show sympathy to us as 'heroin khor'. They permit us to cross the border sometimes'.

Injection of any sort is an even more efficient way of spreading HIV than sexual intercourse. Since injecting drug users are often linked in tight networks and commonly share injecting equipment with other people, HIV can spread very rapidly in these populations. High rates of injecting drug use in Bangladesh have been associated with its proximity to the 'Golden Triangle' where a good quantity of world's heroin is believed to be grown and processed. Being very close to the 'Golden Triangle' and having thousands of kilometres of common border with India, particularly the state of Manipur and West Bengal, good quality heroin is easily available. This illicit drug is fuelling an HIV epidemic among intravenous drug users and is also spreading the potentiality of HIV transmission in Bangladesh. Some aspects of street life such as extreme mobility, low knowledge of HIV and recreational sex compound the vulnerability of street drug users. However, it can be said that drug addicts have to be geographically aware and tactical users of place due to the stigma attached to their activities and the identity crisis that this causes, and, as a result, they try to take drugs in controlled areas where they can manage everything themselves and nobody will disturb them. On the other hand, they do not hesitate to go to any place where they will have access to drugs. Basically, they collect their everyday drugs from local dealers but when there is a supply crisis, a price hike or they need the assurance of a good quality product, then they have to move. The tactics of place selection for drug collection can make the addicts vulnerable. Homelessness is associated with elevated levels of HIV and related risk behaviour among IVDUs, influenced by living conditions and lack of socio-economic resources. In addition, border and urban growth points can increase the health risk, particularly HIV vulnerability, not only through high population movement, the sex and drug trade, and availability, but also due to fear of violence, exploitation or deportation.

Sex Workers' 'Risky' Places Almost all prostitutes, including brothel and non-brothel workers, face a wide range of violence, including the extraction of money, free sex, harassment and arrests, across the full range of different locations. Some examples are in the following.

Girls' Views About 'Safe' Places Violence is a common theme in the prostitute's life. Different girls have different conceptions about the relationship of risk and place. They disagree, for instance, about degrees of safety between a madam's residence and a hotel, but almost all of them consider customers' houses to be unsafe because of the unpredictable attitude of the punter. Some girls think that risk can come from unprotected casual sex with multiple partners, some also consider the place as unsafe. In Jessore and Khulna, there are a few recognized brothels but many unrecognized and hidden brothels in both. For example, many hotels have a sex sideline for the sake of their sustainability. This is an open secret to all. Interestingly, there is another type of brothel which is located in ordinary residential properties and stays hidden. Many sex workers prefer these so-called *madam* houses as the safest places for commercial sex. The *madam* looks after the safety issue and protects them from neighbours' suspicion. Each madam keeps only one or two girls, who are known to her neighbours as her 'daughters'. Sometimes additional girls visit the house after a phone call from the 'madam' when she has a request from a potential customer. The customer and the girl visit the house as her relatives. In order to stay in business, the madam will have to pay a bribe to the police and protection money to local *mastaans*. But if local people hear about it, they will need to move away, to a new place. I asked a girl, Ruma, who is mostly working at the residence under a madam's control, what are the safety issues in the residence. She told me that '*when police sources trace our madam's house, the madam gives them money and a girl, whoever they want; then they become quiet and they protect the madam and her girls*'. On the other hand, some girls told me that hotels are safe and residences are risky after only two to three months when it becomes known and the locals can identify the sex workers easily. Hotels are comparatively safe where there is a good connection between the hotel management and the local administration and so girls are less likely to be victimized. When I asked the girls about the advantages and disadvantages of hotels and residences, one girl, Keya, who works in both told me that '*actually the madam's house or residence is fine; I don't need to mix with many people. The disadvantage is that customer flow is low. On the other side, in hotels there is continuous work and we can earn easily but the problem is that we have to face police raids*'.

Risks in 'Safe' Places Many hotel-based sex workers are obliged to have sex without a condom with known customers and the hotel manager. Sometimes hotel managers request them to have sex without a condom for a 'VIP customer' (an influential customer). In one FGD with hotel girls at Khulna, they said that '*we cannot force about two out of five customers to wear condoms, and some hotel owners request us to take some customers without a condom*'. Sometimes hotel-based sex workers are not fully paid and need to provide money to brokers. There are many opportunities to be cheated, for instance, as a result of the shift systems operated by hotel managers because customers want 'fresh' and 'new' girls for each visit. If a girl does not get a customer in her allocated shift, her income is uncertain and she becomes more dependent upon the customer's choice. She may be pressured into having 'free sex' in order to guarantee another shift in the future. I found one hotel- and residence-based sex worker, Keya, who told me that she has to have sex with hotel managers because '*I have to convince them and need to develop their trust on me. It's in my interest for continuous business*'. Many residence-based sex workers, also considered 'call girls' due to their high status, are fearful of visiting new places. Some go with their '*bandhobi*' (friends) to a madam's house but, until they reach the house, they do not know where they are going. They are always fearful that they will meet someone who will recognize their family identity. However, when I enquired about the health risk situation in residence houses, one girl confessed that sometimes the madams will encourage them to take customers without a condom, suggesting that it is enough to use contraceptive pills and have a thorough wash. Some customers offer 1000 taka for 'one shot' without a condom, and since many sex workers are poor, they may be tempted by 30 minutes of work for such a sum.

'Contract' Trip and Preferences Floating and street sex workers need to go to many uncertain places for their 'business' with the customer, which causes many dangers and physical risks for them. Many floating or casual sex workers travel around the region with customers well known to them and stay with them for few days in some cases. I found a few girls who have had this experience of travelling around the region with their customers, staying in hotels as husband and wife. I also found some girls willing to travel long distances for 'contract work', visiting, say, Dhaka or Faridpur and staying there for up to a week. At Benapole, I found sex workers who go for these 'contracts' and visit many districts. These girls are known as '*saree*' (traditional clothes) sellers in the area. During the 'contract' trip, they have to face rough behaviour and suffering and customers also force upon them 'bad requests'. One girl, Tania, explained that some customers want anal or oral sex and it is dangerous to refuse. When she is on the customer's territory, it is difficult to manage his attitude, especially if he is drunk or refuses to use a condom. A group of floating sex workers in Khulna shared their experiences with me about these 'outside' risks:

> 'If you go outside, you must to face many types of problem. First, you will need to work with many people rather than a few. Second, with people coming and going up to midnight, there is the fear of local people knowing. Then we face many dangers, particularly police harassment and threats of rape. Third, among the customers, if anyone has sex without condom and if he mentions it to others, then they will also want to do it without a condom, and we are without protection'.

A sex worker may also find that she does not have enough condoms if she faces many people beyond her expectation. One NGO counsellor told me that girls working at night in a customer's house may be threatened with a knife so as not to have to use a condom, and sometimes they have to face group anal, oral or vaginal sex. Sometimes they take hard drinks with their clients and as a result are more susceptible to abuse, and sometimes they are not even able to collect money from their customers. Floating and street girls are so marginalized and economically insolvent that when a customer contacts them by mobile phone or on a street corner, they usually bargain about payment and its surety rather than checking the 'danger' of the place of assignation. They think that they will stay there for a few hours, not a whole day and night. One floating sex worker Lota told me that '*I ask the punter about the payment first, then assess the 'danger' of the place where he wants to bring me. If I feel the place would be 'safe' for me I decide to go*'. Regarding the relationship between place and risk, many street sex workers in Benapole say that there is no fixed place for their work. They need to conduct their business in places like on the road, on the hillside, by the river, under a truck, in the bushes or wherever they can find somewhere, so they cannot negotiate about their health protection measures. When they get a customer, they aim to finish the work quickly and do not think so much about the place. Silpi told me that '*we get little chance and we need to finish the work in a very short time, so how can we manage condom use*?' I found some street girls at Khulna railway station who use places such as empty train compartments, under the rolling stock and behind shops, meaning that they cannot properly negotiate about payment and condom use as well. Sometimes, when they are caught sight of by police or guards, they need to negotiate the matter very quickly, sometimes to give a bribe or give their body in order to protect themselves from prison or a mob of local people.

"Brothel Is Less Vulnerable than outside" It is generally considered that a brothel is a risky place for HIV infection or transmission but many brothel girls disagree with this. Their logic is that they are careful and aware about condom use and the brothel itself insists on condom issue. Outside the situation is different because there is no control of condom use. Moreover, many young age girls are having casual sex because of economic hardship. One peer educator, Bobita, who had worked in a brothel as a sex worker for a long time, told me:

> 'On the streets casual girls can be killed or tortured for insisting on a condom but in the brothel there are so many people available here with views about condom use. In the brothel, girls not only use condoms, but also check the men's bodies'.

In the view of many brothel girls, it is the customers who are responsible for the potential spread of the HIV virus. Their lack of care with condoms and their sexual behaviour puts the women in a most vulnerable position. It is well recognized that non-brothel sex workers in Bangladesh comprise a significant mobile population. The level of risk of violence associated with sex work is amplified by the space where the work is taking place. For example, women who work solely on the street indicate much more vulnerability with regard to not being able to assert their power with clients. They experience high levels of physical abuse. On the other hand, those women who work mainly in hotels or in a madam's residence have the added advantage of having in-house security. Many reported that the security was effective and that they felt safe. If they had a problem with a client such as showing threat or unwilling to make full payment, then security was on-hand and was helpful in dealing with the problem. However, many hotel sex workers complained that hotel management and security could get violent if sex workers were unable to pay rents at the end of the month or week, and that they sometimes demanded sexual favours. Besides, hotel- and street-based sex workers suffer violence and abuse from the police and from *mastaans* and the hotel girls are often arrested and harassed. Regarding the brothel as a place for sex workers, it is considered that many sex working women are too powerless in the face of a strong brothel power structure to be able to achieve high levels of condom use through negotiation on a person to person level with a client, though the law and order situation in brothels has been improving in Bangladesh due to an NGO presence in many. Despite this, brothels remain non-secure places for many due to competition among the girls to secure customers, the need for girls to obey *sardarnis* or house owners and facing the police and other influential people. However, due to the closure of city brothels such as Tanbazar, Nimtoli and Magura and an increased demand for sex workers in non-stigmatized locations, there has been a remarkable change in the nature of the non-brothel-based sex work in recent times. Moreover, in slum areas many women live in very difficult conditions in terms of income because they have no regular husband. Many slum girls are married by the age of 12–13 years and within a year they are giving birth to their first baby. Divorces are common and many have problems with their husbands, such as long absences or the husband having two or three wives. For many slum women, prostitution is a coping strategy and may be combined with being a construction worker or a maid servant.

Transport Workers' Preferred 'Sites' Truckers need to move across the country for goods transfers and they stay away from their families for long periods. As a result, many of them have been habituated into commercial sex and alcohol as a part of their leisure and also consider it as stimulating for the driving profession. Many ethnographical literature reviews show that long-distance truckers have been found to participate in vigorous or diverse sexual cultures at roadside settlements and border points with poor young women including commercial sex workers. Truck drivers and their helpers in Bangladesh are at high risk of contracting HIV because of their frequent absences from home and their sexual contacts with prostitutes. Some examples are in the following.

Transport Workers' 'Refreshing' Sites There are some reasons frequently articulated for transport workers having relations with sex workers which are absence from their wives, a desire for fun, peer pressure and importantly, the glamour and sexual techniques of sex workers and their willingness to engage in acts that subjects did not want to perform with their wives. Most Bangladeshi truckers, who visit Dhaka, favour this city as a good place for encounters with sex workers in the street or in congested truck terminal areas. Many Benapole-based truckers prefer to have sex in the Goalondo ferry terminal on the way to Dhaka or the wholesale markets in Dhaka such as Armanitola and Babubazar, during the loading and unloading of goods. Many drivers also prefer either a hotel or a friend's house where they will not be disturbed. Rather than hotel and brothel girls, many transport workers prefer 'site' girls (those who are based in local street and floating). They consider these to be 'safer' (from their point of view) than hotel and brothel girls in terms of the number of customers and cleanliness. When truckers drive during the night, in some remote and isolated areas they get 'signals' by someone spreading lentils on the ground, which means that someone is available. If they want to proceed, then they park the truck beside the road and finish their 'business' in a dark place within a short time. One truck helper, Mahbub, told me about the reason for his preferences:

'I like 'site' girls but not brothel girls because many people go there and brothel girls have sex with many customers. But site girls do sex with only a few men. So, they are satisfied with a few customers and we do not need to use condoms. So, we do it in the open air and that's my preference'. When I asked him about the risk of having sex with 'site' girls, he said *'why should I be afraid, we are men! If the girls can come out from their house at night, so why should I be scared? I am not afraid about that'*. Mahbub also admitted another reason behind choosing 'site' girls: *'There is no hurry with 'site' girls. If there's no-one after me, then I can enjoy whatever I like. It doesn't matter where I am biting or kissing her, if she is good girl, she will allow me to bite or kiss her body'*.

Different Views About 'Safe' Places One helper, Zillur, was opposed to sex at 'sites'. His logic is of dangers on the streets, such as being caught 'in flagrante'. He prefers a '*para*' or brothel because he is more relaxed there. Another transport helper, Sukkur, does not want to be restricted to the *para* although he fears being captured by local people and being considered a 'bad guy'. Some drivers told me that they feel a need for sex but they exercise self-control because, if they go to the local *para*, it might be possible for them to be traced. If they go instead to a hotel, this is less likely. Faruk said that *'hotels are comfortable. Nobody will see you. If you ask the hotel manager he will supply a girl, so I don't need to go to any trouble. This opportunity is not available in the brothel, where girls misbehave with the customers for extra money. But in a hotel you just offer the girl your contract money only'*. However, some drivers with a Muslim sensibility follow the religious dictum to avoid brothels but this does not stop them looking for street girls. However, during an FGD with rickshaw pullers in Khulna, all were agreed that they prefer brothel girls because of their affordability. Many also pick up girls from the park, street or station area who sell sex for a low price. Here price was a priority over 'safety' concerns. When they saved sufficient money, they visited the local brothel named Fultola, which is located close to Khulna city. Some rickshaw pullers consider this brothel to be 'safe' because its environment is good. There is no noise and the brothel is surrounded by a brick wall.

5.3 Mobility and Role of Border Towns

Analysis of the geographical distribution and migration of HIV is an established field of study (Gould, 1993; Smallman-Raynor, Cliff, & Hagget, 1992). Although the cause of the global spread of HIV is complex and multifaceted, increasing population mobility both within and across countries has been implicated as a major factor for spreading the virus and migrants are more vulnerable to HIV than non-migrants (UNAIDS & IOM, 2001). The relationship between population mobility and the spread of infectious diseases in general has been examined in many different contexts around the world (Massey, Arango, et al., 1993). For the acquisition of HIV, movement has been identified as an independent individual risk factor in a wide range of settings (Lagarde et al., 2003). HIV infection is itself a trigger for mobility in a variety of contexts (Berk, Schur, Dunbar, et al., 2003). Smith (2005) adopts a migration streams framework which is shaped primarily by the social networks of individual migrants who play a role in shaping the geography of new HIV infections. It is generally reported that migrant workers in developing countries are more likely than non-migrants to take part in HIV risk-taking behaviours. In other words, rural-urban migrants are more likely to be disease carriers (Castle, 2004), and are frequently identified as 'bridging populations' for HIV transmission (Morris, 1997). Mobile populations including sex workers, drug users and transport workers may themselves be considered as 'core' or as 'bridging populations' (Cates & Dallabetta, 1999) for transmitting HIV/STIs to lower-risk groups in countries/regions of origin as well as destination.

In other words, migrants may act as a bridge population in the spread of HIV as infected migrants return home with the virus and unknowingly pass it onto their sexual partners (Hirsch, Higgins, Bentley, & Nathanson, 2002; Lurie et al., 2003). Numerous studies have cited migration as one of the important factors leading to the rapid diffusion of HIV in the context of developing (Caldwell, Anarfi, & Caldwell, 1997; Campbell & Williams, 1999; Wolffers, Fernandez, Verghis, & Vink, 2002) and developed countries (Gras, Weide, Langendam, Coutinho, & van den Hoek, 1999; Lansky et al., 2000; Wallman, 2001; Wood et al., 2000). Economic marginalization, social isolation and lax social control all contribute to elevated HIV risk behaviours among migrants (Yang, 2006). Ming (2005) also puts HIV/AIDS in the context of cross border mobility. However, borders and major trade routes are physical structural determinants of heightened HIV vulnerability given that they facilitate population movement and mixing. Rhodes et al. (2005) identified critical factors such as border trade and transport links (Lacerda et al., 1997), population movement and mixing (Hammett et al., 2003; Saggurti et al., 2011) and urban or neighbourhood deprivation and disadvantage (Wood et al., 2002) in the social-structural production of HIV risk. In many literature, truck drivers have been identified as having high-risk lifestyles for STD transmission in India, Thailand and sub-Saharan Africa (Podhisita, Wawer, Pramualratana, Kanungsukkasem, & McNamara, 1996; Singh, Sivek, Wagener, Hong Nguyen, & Yu, 1996).

In Bangladesh, it is a great concern that surrounding nations India, Myanmar and Nepal have high rates of HIV and it is commonly thought that HIV could be 'imported' (Chowdhury, Choudhury, & Lazzari, 1995; Gibney, Saquib, & Metzger, 2003), especially from India through cross-border smuggling activities, and the behaviour of Indian truckers and Bangladeshi sex workers. In a study of non-brothel-based female sex workers in India, Dandona, Dandona, Gutierrez, Kumar, et al. (2005) found that street-based workers are a significantly higher HIV infection risk compared to brothel girls due to them being 3.5 times less likely to use condoms with clients. Gangopadhyay, Chanda, Sarkar, Niyogi, et al. (2005) found that in West Bengal, the Indian state bordering Benapole, many sex workers agree to unprotected sex if forced or offered extra money. In a study to assess the West Bengal brothel girls' HIV status, Sarkar et al. (2005) found that the prevalence of HIV infection was 9.6% including 0.6% HIV-2 infection, which probably indicates their sexual exposure to multinational clients, particularly seafarers. With a highly mobile population in aggregate, the border towns may be considered high-risk environments, or even 'core environments' with little differentiation in levels of risk between groups within them. People working here might be considered as 'bridging populations' when they return from the town to their home region. Here, the 'bridging' concept defines the transfer of health risks to what we might call 'innocent' parties. In this study, I found much evidence of this kind of mobility pattern and 'risky' behaviours by local sex workers and Indian truckers at Benapole port near the border.

(In)visible Role of Border Women Involvement in the commercial sex industry is generally characterized by a high level of geographical and occupational mobility. In the border area of Bangladesh, smuggling is a regular business and regarded by many as a legitimate source of earning. I was told that many of the girls involved in smuggling are used as a means of negotiating with the Indian Border Security Force (BSF): sex traded for ease of transferring goods from India to Bangladesh. Basically, Bangladeshi women bring fruit, eggs, banned drugs including heroin, etc. from India and they cross the border with the help of '*ghat*' (illegal crossing points) people. Those who cross the *ghat* illegally have to wait for a signal from the middlemen on both borders, who are called 'the syndicate'. They control the border security forces through bribes and the trafficking of 'women' for sex. The middlemen give a 'token', a special small document which acts as a pass. A smuggler woman Monowara explained this:

> 'When I have money I make a payment to the syndicate people for allowing me to cross the 'ghat'. Otherwise I must have sex with them. I prefer this rather than giving money, for the sake of my livelihood. Sometimes the syndicate people take wine/ alcohol before sex and this can take a long time. Sometimes I need to do sex with Indian businessmen and often also with the Indian BSF when I come back to Bangladesh. They usually do it in the evening in a dark place. After sex I then go to somebody's house in the name of using the toilet, and wash my vagina, otherwise the HIV disease will come, and I will die, so I try to wash'.

Population movement is structurally connected with both economic disadvantage and social inequity. For example, in many poor countries poverty is the primary force driving the migration of women from rural to urban areas, where they are often tempted or absorbed into the sex trade. This reminds us that a number of environmental issues combine to produce geographic effects in HIV association with migrant women. In the context of Bangladesh borders, migration is a central feature of life in the border towns. Large-scale trade between India and Bangladesh has brought a rapid influx of different people including sex workers to the borders. Some migrant women come in search of employment. It is locally known that there are eight *ghats* around Benapole port and its 2–3 mile hinterland. As a result, everyday mobility has been causing many health risks but these are mostly ignored by the girls due to their need for a livelihood. Although these girls have given their 'consent' for sex, most of the time they are forced into sex without taking any protection in the exchange for easy accessibility across the border. Many Bangladeshi truckers believe that brothel girls have the 'germ' but not Benapole's street or smuggling girls.

'OPEN Secret' Role of Indian Truckers Most studies consider migration to be a major vector of HIV transmission, which can be attributed to migrants' risky behaviour. Transport workers who stay long hours on the road and often spend several days in one place for clearing customs or resolving mechanical problems of vehicles seek entertainment through sexual activity mainly as they are away from their spouse or regular sexual partner. In other words, sexual freedom appears to be one of the attractions of the town or a compensation for separation from home to the transport workers. Many drivers have reported a large number of non-regulars, usually commercial, partners where they may not get home for several months. In Benapole, 300–700 Indian trucks and more than 500–700 Bangladeshi trucks arrive every day. Indian trucks stay at Benapole terminal from a few days to weeks due to loading and unloading. In the meantime, the truckers introduce themselves to brokers who have good contacts with the local girls. Although there is a gate to the terminal area, at night the girls can enter the area where the Indian truckers stay. Some of the regular Indian truckers have fixed or known girls. When they arrive at Benapole, they inform a broker and give an advance for the 'business' that is to come at night. Regarding the preferred customer, most girls at Benapole prefer Indian transport workers rather than Bangladeshis. They are said to behave better with the local girls and pay more. Although most Indian transport workers do not use condoms, the girls allow it because either they have a good intimacy or are earning a good amount. Monowara told me that '*foreigners talk sweetly and pay well, including tips, as they like Bangladeshi girls, so they spend their money on us, but the Bangladeshis try to exploit us. Bangladeshi drivers pay at most 50 taka, but foreigners give 100 taka. Some Indian drivers even give us loans, so we don't force them to use condoms*'.

During my field work, I discovered another preference for Indian truckers as customers by the local girls which might be a cause for changing risk. Among those sex workers involved with Indian customers, some expressed a preference for Muslim rather than Hindu drivers. They believe that sex with Hindus amounts to a greater 'sin'. Silpi told me about that '*many Indian drivers are hindu and many girls will have sex with them and do not differentiate between hindu and muslim. I am more careful. I only mix with muslims, because I am a muslim girl. Although many girls do not care about it, I prefer muslims because we have to give an account to Allah. I want to know his name first and then I can judge whether he is muslim or hindu. I have good relations with many Indian muslim drivers. When they come they let me know and then they come to the house at night*'. However, some Indian drivers take a 'wife' in Benapole. They choose a girl who has been economically abandoned and spend as little as possible in monthly costs on her. In exchange for marriage, the girl will get some assurance for getting monthly money, and if payments become irregular or stop, then the girl will consider that her husband is not interested in continuing the relationship. One outreach worker, Jabed, told me the story of a girl named Jamila of Benapole who married an Indian driver but within three months she got married to another man. Other relationships are, however, more stable. For instance, I met one woman, Popy who married an Indian driver and started a family.

Indian Truckers' Sex from Bombay to Benapole Some Indian driver and their assistants consider sexual exposure to multinational clients as a matter of pride. As many Indian transport workers visit Bangladesh borders, like Benapole, it is believed that they enjoy sex with a '*bangla meye*' (Bangladeshi girl). Some consider these girls to be disease free. Some Bengali-speaking Indian truckers shared their experiences in an FGD about the women who are visible in the custom godowns area and truck terminal area of Benapole. They 'use' these girls at night, either in the parking area or in the adjacent houses of the slum. They (Indian truckers) explained their preference: '*Some of us feel very excited to have sex with Bangladeshi women. The rate is cheap here, basically our 100 rupees being the equivalent of 200 taka. Some girls are available for 50 taka or 25 rupees, whereas in India the minimum rate is 50 rupees, so this is a bargain*'. A few Indian truckers who reported having sexual relations in Bangladeshi towns near the borders also had sex in boarding houses or with hotel sex workers. Sometimes some Indian truckers visit the Jessore brothels. During my interviews in Benapole, I met with some Indian truckers in order to discuss their sexual experiences with Benapole girls. One of them, Abdul, related that:

> 'Since 1991 I have been coming to Benapole and in the last 5 years I have been having sex with the local girls. In Benapole, when we are outside the parking area, the girls come at night and call to us. Some Indian drivers go to the nearest slum or to a girl's house. I prefer not to use a condom during intercourse because it is the girl who will benefit, not me. Though I have some itching problems on my penis, when I take medicine from the pharmacy, it gets better. Other than Benapole, I also enjoy hotels in Bombay and parts of Andhra Pradesh. Bombay is costly but Andhra is less expensive and just as good'.

This case study shows how risk can be imported from one place to another. The girls stated that they have regular sexual and emotional relations with Indian transport workers, who are their main customers. This specific connection is likely to be substantially responsible for spreading the threat of HIV/AIDS and other STDs in this area. Because these drivers and helpers also travel to many different cities in India, including Bombay or Andhra Pradesh, they are an important channel for spreading the virus. From this primary 'imported' health risk, the infection has now moved in a secondary wave to the general population of Bangladesh.

5.4 Hidden HIV Cases in Bangladesh?

The spatial dynamics of HIV is described by Patton (1994) and termed 'sexual geography' because 'HIV has achieved its geographic mobility in the bodies of infected people' (Patton, 1994; p. 21; see also Knopp, 1992). The spatial relations of HIV/AIDS as migration, daily paths, social networks and locations have largely been explored through the infected bodies. The global epidemic of HIV/AIDS has come under scrutiny since the 1980s and, particularly, from the beginning of the 1990s. As a topic of research, it has a strong pull because of the heavy toll of morbidity and mortality, and also because of the intellectual challenge of understanding its epidemiology. As the epidemic has unfolded, few regions in the world have remained free of the infection and a complex spatial and socio-cultural 'geography' of many overlapping epidemics has evolved (Clift & Wilkins, 1995). AIDS outbreak may be a consequence of vulnerable of places, sexuality and mobility (Sabatier, 1996). Feldacker et al. (2010) examined how area and individual level risks and HIV status vary in Malawi. They found that area-level factors of HIV contribute individual-level contributions. Cianelli et al. (2013) discussed that due to age and cultural issues HIV risk can increase. Collins et al. (2016) illustrated spaces which can create challenges to accessing HIV services. Essentially the pattern and pace of HIV transmission in particular geographical settings depends upon the interaction of two main factors: firstly, the nature of sexual and injecting drug using cultures, and secondly, the governmental and societal response to combating the threat of HIV/AIDS (Ford, Siregar, Ngatimin, & Maidin, 1997).

Within South Asia, trafficking of women for the sex trade is common in Nepal, Bangladesh, Pakistan and India (Fernando, 1997). According to one report, every day 50 Bangladeshi girls are lured across the Indian border and sold (Ahmed, 2006). Apart from trafficking for sexual purposes, women in the region are also traded as marriage partners, domestic workers, child-carers, construction workers, beggars, casual industrial workers and nurses, particularly to Middle Eastern countries (Murthy & Sankaran, 2003; Paul & Hasnath, 2000). However, in South Asia, the dramatic rise of HIV infection in India and Myanmar alarmed public health officials in Bangladesh (Chowdhury et al., 1995). Sarkar et al. (2005) found that the overall HIV sero-prevalence among IVDUs in the Indian state of West Bengal is rising and is expected to increase further due to several factors including risky sexual behaviours. This rise in cases, coupled with the rapid spread of HIV/AIDS in India, has raised fears for Bangladesh because of its illustration of how rapidly a minor rate of HIV infection can turn into an epidemic in a country with socio-cultural and economic similarities. Although the rate of infection of HIV is lower than expected, Rahman and his colleagues (1999) think that low HIV prevalence as currently reported does not reflect the true situation. The first HIV case was detected in Bangladesh in 1989 (NASP, 2004) and the rate of HIV infection is predicted to increase in Bangladesh because all of the behavioural and bio-medical risk factors are prevailing in the country (Gibney et al., 2003). In this section, I will look at the issue of women trafficking which is contributing to the spread of HIV infection in Bangladesh. However, the research also seeks the 'unrecognized source' of HIV infection which is playing an unseen role in this conservative society.

'Contribution' from Trafficking Women trafficking issues are very common in South Asia, especially Bangladesh, for the purpose of 'sex business' like the global trends. Several thousand women and girls are trafficked annually from Bangladesh for the purpose of sexual exploitation, primarily to India, Pakistan and the Middle East. Although reports and studies identified these border routes, traffickers use different unusual routes at different times in order to avoid police and other law enforcement agencies. To enter India through West Bengal, the two most common routes are the border in Jessore and Satkhira, two frontier districts of Bangladesh. It is worth noting that a good number of women from Bangladesh have crossed the border into India and engaged in different informal activities like domestic workers, construction workers, etc. in the different cities of India including Kolkata. It is evident that most of trafficked women engaged in these activities are forced to engage in prostitution, especially in Mumbai.

Some trafficked girls are taken to Bombay as West Bengalis. Some start their work there as maid servants but after a time they turn to prostitution. Many are young women from remote villages and poor border communities who are lured from their villages by local recruiters, relatives or neighbours promising jobs or marriage. Many reports show that most of the trafficked women are sold for very small amounts to brokers who deliver them to brothel owners in India for anywhere from Rs 15,000 to Rs 40,000. This purchase price becomes the 'debt' that the women must work to pay off—a process that can stretch on indefinitely. One Bangladeshi transport worker who has visited many parts of India told me that Bangladeshi women are working in many cities of India including Bombay and Kolkata either as dancers or part-time sex workers. Many of the women working in the brothels of Jessore and Khulna regularly cross the Indian border and spend time for shopping and also working in the brothels in Kolkata, before returning to their previous work. Bangladeshi young women are in demand in India: they look conservative, reserved, over-dressed and '*neerog*' (disease-free), and, in the imaginations of the customers, possibly virginal. Many of the border villages are said to be '*bombay para*' (village named Bombay) because many of the local girls have been trafficked at one time or another to Bombay and some of them have returned with a good amount of money. This encourages other girls' families to fall into the trap of traffickers. During my field work in Benapole, I found one girl, Sajeda, who had recently come from Bombay after a long time and was now doing commercial sex work in the Jessore region:

> 'As we were poor, I went to Bombay from Benapole at the age of 12–13 with a broker, to help my mother and younger sisters. He said he would find me a good job there. This turned out to be a job in a bar. I had sex with the customers without a condom and took many risks. I worked there as a prostitute for three years and earned lots of money. I faced many types of customers who came from different places. I also had group sex experiences with 4–5 people. I came back to Benapole in 2005 because the Indian government forced us to leave their country before the elections. Now I am working here as a sex worker in the Jessore area but sometimes I go to Dhaka with customers. Here customers also do not want to use condoms. I don't know what kind of problems may arise from having sex without condom. I've heard about AIDS, but I am not sure whether it is in my body or not'.

Trafficked girls who become prostitutes have no family trace to report to on the girl's welfare. The trafficked girls forced into the brothels do not want to return to their homes once they are into it for more than one year. Such girls believe they would be victim of social stigma and face discrimination from society and that their families would also suffer. One boundary is the *Isamoti* river but in some places this has no existence in summer and winter and people can go to India by jumping over the channel. The BDR and BSF are unable to trace the traffickers because the people who live along the border on both sides are tightly knit. As the trafficked girls from Bangladesh are undocumented, poor command on Hindi language and no legal knowledge of their rights, so they are vulnerable for sexual exploitations and violence to the border employees in both parts along with the 'border traffickers'.

In West Bengal, Kolkata is considered a hub for the trafficking of women and girls where large numbers are smuggled in from Nepal, Bangladesh or Burma. From Kolkata, they are sold on to brothels in Mumbai and Delhi. In the south western part of Bangladesh, particularly in the Khulna region, a good number of women who have been trafficked back and forward across the border have been identified as PLWH and the 'victims' of women's marginalization. One girl in Khulna named Shahana is a good example:

> 'I lost my jute mill worker father at the age of 6 months along with my brother and two sisters, and my mother spent all of our deposited money on his treatment. Then my mother went to my father's village, but they refused to help, and she came back to Khulna again. Then some people advised my mother to go to India for work. When I was 3 years old, we went to India. My mother arranged my marriage at the age of 12 in Kolkata, and I became a mother at 14. My husband was a labour-cum-contractor but he didn't contribute to the family expenditure. So, I took work, as an assistant to a carpenter but I left due to sexual harassment. I couldn't manage to give a proper diet to my small daughter, then I started a little vegetable business but some mastaans disturbed me and threatened rape. I tried for many jobs, including maid servant. In the meantime, I gave birth to my second baby and my husband died accidentally in Delhi. Then one of my relatives advised me to go to Bombay as I was workless. I left my babies with my mother but I didn't talk about my expected profession. I promised to send her money. I decided to go to a Bombay brothel for the sake of my children's future. At the beginning I was busy but within one month I become pregnant and had to come back to Kolkata for an abortion. My mother understood when she saw my symptom - vomiting and weakness - and she told me to beg rather than do sex work. But I was determined to go there again in order to give an education to my babies. I saved a large amount of money, but after some time, I had some physical problems, like frequent fevers. Most customers didn't want to use a condom and we were not allowed to refuse a customer. I visited doctors often and was cured. After two and a half years I gave my savings to my brothers to buy a land in Bangladesh. Then I met a Bangladeshi man who wanted to marry me. One day I went to a medical centre in Bombay for a blood test and they asked me to collect the report in the evening. When I went there in the evening I was surprised to see that everyone was looking at me. I didn't understand what was going on until they told me that HIV had entered my body'.

Many people believe that many women migrate from one country to another or from rural to urban areas by choose through the help of traffickers. In this case study, Shahana also accepted her profession knowingly and gave her 'consent' to start it. However, I found another woman in Khulna whose whole family had migrated to India to a village in the state of Karnataka. Her family arranged her marriage with a local villager who was a lorry driver. After her husband's death, one of her babies also died due to a lack of proper diagnosis and treatment. Then she, along with her daughter and other family members, returned to Bangladesh and a few years later she and her daughter were identified as PLWH. She believes that her present disease came from her husband because there were no other possible sources of infection.

Unidentified 'Sources' A good number of Bangladeshi people travel frequently to some countries which are known to have high HIV prevalence, such as Thailand or India, for leisure, sightseeing, medical treatment, business, etc. In addition, about two million Bangladeshi people currently work in different countries, including UAE, Saudi Arabia, Malaysia and Singapore, have an insufficient AIDS awareness. Many of the overseas workers are completely lacking any knowledge of AIDS which ultimately fuel the HIV vulnerability in Bangladesh due to their high-risk behaviours in the foreign countries. In this section of the chapter, I will focus on some participants who are already infected. Some are infected as a result of their risky behaviour but some are 'innocent victims' who do not know the source of their infection.

Case 1 In Khulna, I found one PLWH named Shahana who was infected in a Bombay brothel, who told me that in many Indian brothels they have no opportunity to choose their customers and they have to work with everyone without protection. According to brothel rules, if they refuse a customer, they face physical abuse. Most trafficked girls and women are sexually abused; often they also experience other forms of physical and psychological violence. They are in the highest category of risk of becoming infected with STDs, including HIV/AIDS. Shahana said that she met many Bangladeshi men in Bombay. They visited the brothel and then returned to Bangladesh and slept with their wives. For example, a PLWH has died in Noakhali General Hospital who worked in India for a long time as a labourer. Shahana believes that many hidden PLWH cases like her are living in Bangladesh.

Case 2 There are many Bangladeshi girls who work in the garment industry, fish companies or in construction work for their survival. In Khulna, I found one PLWH named Keya who was involved in her life with all three and was infected with HIV. She started her life as garment worker in Chittagong and then joined a fish farm in Khulna as a helper. She was also involved in '*jogali*' (construction) work during off season at the fish farm. As her husband did not like her fish farm work, she switched to *jogali* work in the Khulna region. After leaving her husband, she went to India for few months as a dancer and bar worker. One of her 'close friends' sold her to the Baniashanta brothel and she started her life as sex worker. In the Baniashanta brothel, a brothel located beside the sea port of Mongla, many girls have experienced sex with foreign ships' crews, mainly from China, Korea, Burma, Thailand and Greece. Although they paid well, they never used condoms. After a while Keya developed pains in her wrist and ankle, and she sometimes had a fever. She took medication but was not cured. Finally, she left the brothel and returned to *jogali* work. Here the female workers are forced to have unprotected sex with the '*mistri*' (contractors), otherwise they are not given work. Meanwhile, her physical sufferings continued. The doctors thought she had rheumatism but then she developed gangrene problem in her leg and was admitted to hospital where her blood was tested and she was found to be HIV positive. There is a strong possibility that Keya has already transmitted her virus to many people who are still unidentified. From this case study, it seems that there are some important considerations about Keya as an invisible pathway of transmission. Firstly, if she caught the virus from a *jogali* worker before she joined the brothel, she must have transmitted it to many others. Secondly, if she caught it in the brothel from foreign sailors, there would also have been many possibilities of transmitting it to many local customers in the brothel at that time and, in addition, to many other people when she left the brothel and worked again as a *jogali* worker until her diagnosis.

Case 3 For migrant men away from the conservative culture of home they feel less inhibited and, with money in their pocket, they are vulnerable. Afsar spent a few years in Malaysia as a migrant worker and shared with me his experiences about the risks of getting HIV in Malaysia. He thinks that nobody would consider Malaysia a Muslim country if they saw the availability of foreign girls, especially Chinese and Russians, who enter on tourist visas. These girls have a contract with a hotel through an 'agent'. Usually they rent some houses and their agents provide the customers, and it is a very profitable business. Among the foreign girls, Russians are most in demand among the migrant workers. Afsar talks about how he became infected:

> 'When I went to Malaysia, I was only 18–19 years old. When people leave the country, and they have no guardian over them, there is scope to make mistakes. In my life I made some mistakes. I don't know where I got the virus. I think my behaviour was responsible for that because I had many sex relations'.

Case 4 Migrants have no political rights and they are economically under-privileged and socially displaced in the receiving countries. In most receiving countries, migrants are subjected to mandatory health tests only to ensure that they will not infect local people with disease. Their health is screened without prior explanation of what they are tested for and how they will be tested. Without pre- and post-test counselling, even if migrant workers are found to be PLWH, they would not know their sero-status and thus would be denied the opportunity to seek treatment at an early stage. Citing national health security reasons, receiving countries usually deport PLWH migrant workers. Uninformed PLWH migrant workers are sent home ignorant of the risk of spreading the disease back home.

In Bangladesh, it is believed that migrants carry the virus from abroad, particularly when they return from periods of work in the Middle East. An irony of HIV in Bangladesh is that wives of these migrants are the most numerous group of the infected. I found one PLWH woman in Khulna who told me that, despite her honesty and dedication to a 'pure' unmarried life, she got HIV as soon as she got married. She had grown up in a conservative family, so she had no sexual experience. But her husband was a migrant worker in Saudi Arabia with experience of commercial sex. She does not know whether the HIV he passed on to her was from abroad or from a local brothel.

Case 5 Recently, one of my interviewees named Salma who recently identified as PLWH told me that she would be reluctant about insisting upon condoms because if she asked a 'friend' to use a condom, he would be curious about her motives and she might be caught and even killed for transmitting the virus. This is despite the possibility that she might transmit the virus to a new group of partners. She stated:

> 'I want to make family life again, but how? I am thinking that I will go with men but ask them to use a condom, though it's a difficult matter. But if anybody wants to sleep with me and I tell him to use a condom, a problem may arise. If I try to protect him, he might suspect that I have a problem. And if he doesn't want to use a condom, what can I do? I will not tell him about my HIV status'.

Sex may be an emotive subject for antibody positives or those who have AIDS. Salma has not had sex with anyone since she was diagnosed. She says that this is because she has not met anyone she wants to have sex with. But she also recognizes that having the virus has made her stop and think a lot more about it. A woman who has AIDS may use help and support in telling her potential sexual partner although it would be a very difficult task.

Case 6 Like other parts of Bangladesh, in Khulna there are foreigners involved in construction projects such as highway bridges and communications projects. In the case of construction workers, many marry local girls on a contract basis. For example, during the Rupsha Bridge construction, foreign workers gave money to the bride's family for this kind of arrangement. I talked with a few girls who had sexual experiences with foreigners. One of the hotel-based sex workers of Khulna named Papri told me that she had experience with two foreigners.

Case 7 Other than sexual contact and needle sharing, blood transfusion is the next most serious risk factor for HIV transmission. Like many developing countries, in the Indian sub-continent blood transfusion is still largely unscreened. I found one PLWH in Khulna, Saiful, who had been diagnosed very recently along with his wife. Saiful got married at a very early age and had their first baby terminated because they were then still teenagers and poor. Two or three years ago when his wife had anaemia due to a miscarriage, he followed a doctor's prescription and bought a bag of blood from a private blood bank and gave it to her in a private hospital. They were shocked when HIV was identified in both their bodies.

All of the above examples show that HIV is spreading in Bangladesh silently and invisibly. Generally, most cases of HIV in Bangladesh are being discovered by accident. The subjects are tested for another disease and are detected as positive. Bangladesh has high rates of internal and external migration. Some of the male internal migrants (from rural to urban areas) who live in the cities share the risk with their wives or other sexual partners when they return to their home villages. External migrants, other than illegal migrants to India, go to the Gulf States or South East Asian countries. Again, many are at risk of HIV/AIDS because of casual sexual relations with foreign commercial sex workers. If they are identified as PLWH they may be forcibly returned home. People returning home with the virus may feel ashamed to be tested for fear of being socially isolated. In Bangladesh, there is no arrangement for the migrant workers about voluntary counselling on HIV transmission, and this is urgently required. It is especially timely to call for public initiatives to provide AIDS information to migrants, particularly overseas job seekers before they leave Bangladesh.

5.5 Place, Mobility and HIV

Much empirical research shows a connection between place, mobility and HIV spread. This research also shows how geographic place can be a contributing factor for health risks particularly HIV infection for vulnerable people in the context of Bangladesh. Among the 'most at-risk' group, sex workers can be victimized due to their interaction with 'risky' places. Drug users also occupy risky environments in the form of 'shooting galleries'. In addition, transport workers' preferred 'safe' places also put them into vulnerability of infection. Apart from common 'most at-risk' group in the country, widespread illegal migration and trafficking of women in the border areas as well as foreign transport workers play a large part in channelling the HIV risk from neighbouring countries especially India.

Risk behaviours, however, are not fixed in time and space. There has always been an important relationship between health risk and geographic space, particularly in border areas. Location is a factor in HIV infection through high-risk sexual activity, as proven in much geographical research on HIV. But to date there has been no significant empirical research in Bangladesh on the importance of location or the role of place in examining exposure to potential health risk, particularly HIV. Having said that, Bangladesh policy planners have commented on the need to include geographically significant areas such as border towns, where a neighbouring country's citizens are present, in surveillance systems. But, so far, they have not included these places or vulnerable people in specific programmes. No official studies have been conducted to examine these issues, although academic researchers such as Laura Gibney, Bruce Caldwell and their collaborators have given prominence to the border towns. The theory that HIV could be imported into Bangladesh through risky cross border contacts, particularly through (in)visible relationships between sex workers and truck drivers, is borne out through the present empirical fieldwork. There is strong evidence that migrants act as bridging groups. As many Indian truckers need to stay for longer in the many land ports of Bangladesh, particularly at Benapole port for customs checks, so there is a high possibility of transferring the health risk to the Bangladesh border women and to their sex partners back home. This situation is exacerbated by the inability or unwillingness of most of the border women to take any precautionary measures to protect themselves. Moreover, 'risky', and 'preferring' sites or places are also of critical importance in spreading the virus across the country.

The present research has found value in an environmental and geographical approach to sexual risk rather than attributing risk solely to individuals. In this case, the poverty of surrounding areas of border towns contrasted sharply with the potential income-earning and social advancement opportunities of the port, leading to an environment of high numbers and turnover of sexual partnerships. The next chapter will look at issues which have been introduced in previous chapters. It will discuss policy issues about the marginalization and stigmatization of sex workers, drug users and PLWH, as well as the importance of risk behaviour and place in the spread of HIV in Bangladesh.

References

Aase, A., & Agyei-Mensah, S. (2005). HIV/AIDS in sub-Saharan Africa: Geographical perspectives. *Norwegian Journal of Geography, 59*(1), 1–5. https://doi.org/10.1080/00291950510020484

Ahmed, H. S. (2006). Facing an uncertain future, Star weekend magazine, *The Daily Star, 5* (121), November 24, Dhaka.

Andrews, G. J., & Evans, J. (2008). Understanding the reproduction of health care: Towards geographies in health care work. *Progress in Human Geography, 32*(6), 759–780.

Andrews, G. J., Cutchin, M., McCracken, K., Phillips, D. R., & Wiles, J. (2007). Geographical gerontology: The constitution of a discipline. *Social Science and Medicine, 65*(1), 151–168.

Berk, M. L., Schur, C. L., Dunbar, J. L., et al. (2003). Migration among persons living with HIV. *Social Science and Medicine, 57*(6), 1091–1097.

Bourgois, P. (1998). The moral economies of homeless heroin addicts: Confronting ethnography and HIV risk and everyday violence in San Francisco shooting encampments. *Substance Use and Misuse, 33*, 2323–2351.

Brown, M. (1995). Ironies of distance: An ongoing critique of the geographies of AIDS. *Environment and Planning D: Society and Space, 13*(2), 159–183.

Caldwell, J. C., Anarfi, J. K., & Caldwell, P. (1997). Mobility, migration, sex, STDs, and AIDS: An essay on sub-Saharan Africa with other parallels. In G. Herdt (Ed.), *Sexual cultures and migration in the era of AIDS: Anthropological and demographic perspectives* (pp. 41–54). New York, NY: Oxford University Press.

Campbell, C., & Williams, B. (1999). Beyond the biomedical and behavioural: Towards an integrated approach to HIV prevention in the Southern African mining industry. *Social Science and Medicine, 48*(11), 1625–1639.

Carlson, R. G. (2000). Shooting galleries, dope houses and injection doctors: Examining the social ecology of HIV risk behaviours among drug injectors in Dayton, Ohio. *Human Organization, 59*(3), 325–333.

Castle, S. (2004). Rural children's attitudes to people with HIV/AIDS in Mali: The causes of stigma. *Culture Health & Sexuality, 6*(1), 1–18.

Cates, W., & Dallabetta, G. (1999). The staying power of sexually transmitted diseases. *The Lancet, 354*(S4), 2.

Chowdhury, A. Q. M. B., Choudhury, M. R., & Lazzari, S. (1995). *Responding to HIV/AIDS in Bangladesh*. Dhaka, Bangladesh: National AIDS Committee, September, Dhaka.

Cianelli, R., Villegas, N., Lawson, S., Ferrer, L., Kaelber, L., Peragallo, N., et al. (2013). Unique factors that place older Hispanic women at risk for HIV: Intimate partner violence, machismo, and marianismo. *Journal of the Association of Nurses in AIDS care, 24*(4), 341–354.

Clift, S., & Wilkins, J. (1995). Travel, sexual behaviour and gay men. In P. Aggleton, P. Davies, & G. Hart (Eds.), *AIDS: Safety, sexuality and risk*. London, UK: Taylor & Francis.

Collins, A. B., Parashar, S., Closson, K., Turje, R. B., Strike, C., & McNeil, R. (2016). Navigating identity, territorial stigma, and HIV care services in Vancouver, Canada: A qualitative study. *Health and Place, 40*(July), 169–177.

Craddock, S. (2000). Disease, social identity, and risk: Rethinking the geography of AIDS. *Transaction of the Institute of British Geographers, 25*(2), 153–168.

Cummins, S., Curtis, S., Diez-Roux, A. V., & Macintyre, S. (2007). Understanding and representing 'place' in health research: A relational approach. *Social Science and Medicine, 65*(9), 1825–1838.

Curtis, S. (2004). *Health and Inequality: Geographical perspectives*. London, UK: Sage.

Cutchin, M. P. (2007). The need for the "new health geography" in epidemiologic studies of environment and health. *Health & Place, 13*(3), 725–742.

Dandona, R., Dandona, L., Gutierrez, J. P., Kumar, A. G., et al. (2005). High risk of HIV in non-brothel based female sex workers in India. *BMC Public Health, 5*, 87.

Del Casino Jr., V. J. (2004). (Re)placing health and health care: Mapping the competing discourses and practices of 'traditional' and 'modern'. *Thai Medicine, Health & Place, 10*(1), 59–73.

Des Jarlais, D. C., & Friedman, S. R. (1990). Shooting galleries and AIDS: Infection probabilities and 'tough' policies. *American Journal of Public Health, 80*(2), 142–144.

Dorn, M., & Laws, G. (1994). Social theory, body politics, and medical geography: Extending Kearns's invitation. *The Professional Geographer, 46*(l), 106–110.

Dunn, J. R., & Cummins, S. (2007). Placing health in context, In Editorial. *Social Science and Medicine, 65*(9), 1821–1824.

Dyck, I., & Dossa, P. (2007). Place, health and home: Gender and migration in the constitution of healthy space. *Health & Place, 13*(3), 691–701.

Elliott, S. J., & Gillie, J. (1998). Moving experiences: A qualitative analysis of health and migration. *Health & Place, 4*(4), 327–339.

Eyles, J. (1985). *Senses of place*. Warrington, UK: Silverbrook Press.

Feldacker, C., Emch, M., & Ennett, S. (2010). The who and where of HIV in rural Malawi: Exploring the effects of person and place on individual HIV status. *Health and Place, 16*(5), 996–1006.

Ferguson, A. G., & Morris, C. N. (2007). Mapping transactional sex on the northern corridor highway in Kenya. *Health & Place, 13*(2), 504–519.

Fernando, N. (1997). *Trafficking in Asia: An overview*. Bangkok, Thailand: Asian Human Rights Commission.

Ford, N., Siregar, K., Ngatimin, R., & Maidin, A. (1997). The hidden dimension: Sexuality and responding to the threat of HIV/AIDS in South Sulawesi, Indonesia. *Health & Place, 3*(4), 249–258.

Gangopadhyay, D. N., Chanda, M., Sarkar, K., Niyogi, S. K., et al. (2005). Evaluation of sexually transmitted diseases/human immunodeficiency virus intervention programs for sex workers in Calcutta, India. *Sexually Transmitted Diseases, 32*(11), 680–684.

Gatrell, A. C. (2002). *Geographies of health: An introduction*. Oxford, UK: Blackwell.

Gesler, W. M. (1998). Bath's reputation as a healing place. In R. A. Kearns & W. M. Gesler (Eds.), *Putting health into place: Landscape, identity, and well-being* (pp. 17–35). Syracuse, NY: Syracuse University Press.

Gesler, W. M. (2005). Therapeutic landscapes: An evolving theme. *Health & Place, 11*(4), 295–297.

Gibney, L., Saquib, N., & Metzger, J. (2003). Behavioural risk factors for STD/HIV transmission in Bangladesh's trucking industry. *Social Science and Medicine, 56*(7), 1411–1424.

Goldenberg, S. M., Strathdee, S. A., Gallardo, M., Nguyen, L., Lozada, R., Semple, S. J., et al. (2011). How important are venue-based HIV risks among male clients of female sex workers? A mixed methods analysis of the risk environment in nightlife venues in Tijuana, Mexico. *Health and Place, 17*(3), 748–756.

Gould, P. (1993). *The slow plague: A geography of the AIDS pandemic*. Oxford, UK: Blackwell.

Gras, M. J., Weide, J. F., Langendam, M. W., Coutinho, R. A., & van den Hoek, A. (1999). HIV-prevalence, sexual risk behaviour and sexual mixing patterns among migrants in Amsterdam, the Netherlands. *AIDS, 13*, 1953–1962.

Gutierrez-Garza, A. (2013). *The everyday moralities of migrant women: Life and labour of Latin American domestic and sex workers in London*. PhD thesis, London, UK: The London School of Economics and Political Science (LSE).

Hammett, T. M., Des Jarlais, D. C., Liu, W., Ngu, D., Tung, N. D., Hoang, T. V., et al. (2003). Development and implementation of a cross-border HIV prevention intervention for injection drug users in Ning Ming County (Guangxi Province), China and Lang Son Province, Vietnam. *International Journal of Drug Policy, 14*(5-6), 389–398.

Hirsch, J. S., Higgins, J., Bentley, M. E., & Nathanson, C. A. (2002). The social constructions of sexuality: Marital infidelity and sexually transmitted disease -HIV risk in a Mexican migrant community. *American Journal of Public Health, 92*(8), 1227–1237.

Hossain, M. R. (2007). AIDS: Epidemic bell ringing. *The Daily Star*, October 28, Dhaka.

Ivsins, A., Benoit, C., Kobayashi, K., & Boyd, S. (2019). From risky places to safe spaces: Re-assembling spaces and places in Vancouver's Downtown Eastside. *Health and Place, 59*(September), 102164.

Jackson, D. J., Rakwar, J. P., Richardson, B. A., et al. (1997). Decreased incidence of sexually transmitted diseases among trucking company workers in Kenya: Results of a behavioural risk-reduction programme. *AIDS, 11*(7), 903–909.

Jennings, J. M., Woods, S. E., & Curriero, F. C. (2013). The spatial and temporal association of neighborhood drug markets and rates of sexually transmitted infections in an urban setting. *Health and Place, 23*(September), 128–137.

Kaley, A., Hatton, C., & Milligan, C. (2019). Health geography and the 'performative' turn: Making space for the audio-visual in ethnographic health research. *Health and Place, 60*(November), 102210.

Kearns, R. A. (1997). Narrative and metaphor in health geographics. *Progress in Human Geography, 21*(2), 269–277.

Kearns, R. A. (1998). Going it alone: Place, identity, and community resistance to health reforms in Hokianga, New Zealand. In R. A. Kearns & W. M. Gesler (Eds.), *Putting health into place: Landscape, identity, and well-being* (pp. 226–247). Syracuse, NY: Syracuse University Press.

Kearns, R. A., & Gesler, W. M. (1998). *Putting health into place: Landscape, identity and well-being*. Syracuse, NY: Syracuse University Press.

Kearns, R. A., & Moon, G. (2002). From medical to health geography: Novelty, place and theory after a decade of change. *Progress in Human Geography, 26*(5), 605–625.

Kearns, R., & Milligan, C. (2020). Placing therapeutic landscape as theoretical development in Health & Place. *Health and Place, 61*(January), 102224.

Kesby, M. (2000). Participatory diagramming: Deploying qualitative methods through an action research epistemology. *Area, 32*(4), 423–435.

Knopp, L. (1992). Sexuality and the spatial dynamics of capitalism. *Environment and Planning D: Society and Space, 10*, 651–669.

Kuhanen, J. (2010). Sexualised space, sexual networking & the emergence of AIDS in Rakai, Uganda. *Health and Place, 16*(2), 226–235.

Lacerda, R., Gravato, N., McFarland, W., Rutherford, G., Iskrant, K., Stall, R., et al. (1997). Truck drivers in Brazil: Prevalence of HIV and other sexually transmitted diseases, risk behavior and potential for spread of infection. *AIDS, 11*(Suppl. 1), S15–S19.

Lagarde, E. M., Schim van der Loeff, C., Enel, B., Holmgren, R., Dray-Spira, G., Pison, J. P., et al. (2003). Mobility and the spread of human immunodeficiency virus into rural areas of West Africa. *International Journal of Epidemiology, 32*(5), 744–752.

Lansky, A., Nakashima, A. K., Diaz, T., Fann, S. A., Conti, L., Herr, M., et al. (2000). Human immunodeficiency virus infection in rural areas and small cities of the southeast: Contributions of migration and behavior. *The Journal of Rural Health, 16*(1), 20–30.

Lorimer, H. (2019). Dear departed: Writing the lifeworlds of place. *Transactions of the Institute of British Geographers, 44*(2), 331–345.

Lurie, M., Williams, B., Suma, K., Mkaya-Mwamburi, D., Garnett, G., Sturm, A., et al. (2003). The impact of migration on HIV-1 transmission in South Africa: A study of migrant and non-migrant men and their partners. *Sexually Transmitted Diseases, 30*, 149–156.

Maher, L., & Dixon, D. (1999). Policing and public health: Law enforcement and harm minimization in a street-level drug market. *British Journal of Criminology, 39*(4), 488–511.

Marshall, A.-M. (2017). *Confronting sexual harassment: The law and politics of everyday life*. London, UK: Routledge.

Marshall, B. D. L., Kerr, T., Shoveller, J. A., Patterson, T. L., Buxton, J. A., & Wood, E. (2009). Homelessness and unstable housing associated with an increased risk of HIV and STI transmission among street-involved youth. *Health and Place, 15*(3), 783–790.

Mass, B., Fairbairn, N., Kerr, T., Li, K., Montaner, J. S. G., & Wood, E. (2007). Neighborhood and HIV infection among IDU: Place of residence independently predicts HIV infection among a cohort of injection drug users. *Health and Place, 13*(2), 432–439.

Massey, D. (1999). Spaces of politics. In D. Massey, J. Allen, & P. Sarre (Eds.), *Human geography today* (pp. 279–294). Cambridge, UK: Polity Press.

Massey, D. S., Arango, J., et al. (1993). Theories of international migration: A review and appraisal. *Population and Development Review, 19*(3), 431–466.

Mayer, J. D. (2005). The geographical understanding of HIV/AIDS in sub-Saharan Africa. *Norwegian Journal of Geography, 59*, 6–13.

McCreanor, T., Penney, L., Jensen, V., Witten, K., Kearns, R., & Barnes, H. M. (2006). 'This is like my comfort zone': Senses of place and belonging within Oruamo/Beachhaven, New Zealand. *New Zealand Geographer, 62*(3), 196–207.

McNeil, R., Shannon, K., Shaver, L., Kerr, T., & Small, W. (2014). Negotiating place and gendered violence in Canada's largest open drug scene. *International Journal of Drug Policy, 25*(3), 608–615.

Ming, K. D. (2005). Cross-border 'traffic': Stories of dangerous victims, pure whores and HIV/AIDS in the experiences of mainland female sex workers in Hong Kong. *Asia Pacific Viewpoint, 46*(1), 35–48.

Mohan, J. (2000). Geography of health and health care. In R. J. Johnston, D. Gregory, G. Pratt, & M. Watts (Eds.), *The dictionary of human geography* (4th ed., pp. 330–332). Oxford, UK: Blackwell.

Morris, M. (1997). Sexual networks and HIV. *AIDS, 11*(Suppl. A), S209–S216.

Morris, M., Podhisita, C., Wawer, M. J., & Handcock, M. S. (1996). Bridge populations in the spread of HIV/AIDS in Thailand. *AIDS, 10*(11), 1265–1271.

Murthy, R. K., & Sankaran, L. (2003). *Denial and distress: Gender poverty and human rights in Asia*. London: Zed Books.

NASP. (2004). *Bangladesh country profile on HIV and AIDS, National AIDS/STD Programme (NASP), Ministry of Health and Family welfare*. Dhaka, Bangladesh: Government of Bangladesh.

O'Neill, M. (1996). Researching prostitution and violence: Towards a feminists praxis. In M. Hester, L. Kelly, & J. Radford (Eds.), *Women, violence and male power: Feminist activism, research and practice*. Buckingham, UK: Open University Press.

Oppong, J. R. (1998). A vulnerability interpretation of the geography of HIV/AIDS in Ghana, 1986–1995. *The Professional Geographer, 50*(4), 437–448.

Orubuloye, I. O., Caldwell, P., & Caldwell, J. C. (1993). The role of high-risk occupations in the spread of AIDS: Truck drivers and itinerant market women in Nigeria. *International Family Planning Perspectives, 19*(2), 43–71.

Parkin, S., & Coomber, R. (2011). Public injecting drug use and the social production of harmful practice in high-rise tower blocks (London, UK): A Lefebvrian analysis. *Health and Place, 17*(3), 717–726.

Parr, H. (2002). Medical geography: Diagnosing the body in medical and health geography, 1999–2000. *Progress in Human Geography, 26*(2), 240–251.

Parr, H. (2004). Medical geography: Critical medical and health geography? *Progress in Human Geography, 28*(2), 246–257.

Patton, C. (1994). *Last Served? Gendering the HIV pandemic*. London, UK: Taylor & Francis.

Paul, A., Atkins, P. J., & Dunn, C. E. (2012). Borders and HIV risk: A qualitative investigation in Bangladesh. *Oriental Geographer, 53*(1), 73–82.

Paul, B. K., & Hasnath, S. A. (2000). Trafficking in Bangladeshi women and girls. *Geographical Review, 90*(2), 268–276.

Philo, C. (2000). The birth of the clinic: An unknown work in medical geography. *Area, 32*(1), 11–19.

Phipps, A., Ringrose, J., Renold, E., & Jackson, C. (2018). Rape culture, lad culture and everyday sexism: Researching, conceptualizing and politicizing new mediations of gender and sexual violence. *Journal of Gender Studies, 27*(1), 1–8.

Podhisita, C., Wawer, J. M., Pramualratana, A., Kanungsukkasem, U., & McNamara, R. (1996). Multiple sexual partners and condom use among long-distance truck drivers in Thailand. *AIDS Education and Prevention, 8*(6), 490–498.

Rahman, M., Shimu, T. A., Fukui, T., Shimbo, T., & Yamamoto, W. (1999). Knowledge, attitudes, beliefs and practices about HIV/AIDS among the overseas job seekers in Bangladesh. *Public Health, 113*, 35–38.

Rao, K. S., Pilli, R. D., Rao, A. S., & Chalam, P. S. (1999). Sexual life style of long distance lorry drivers in India: Questionnaire survey. *British Medical Journal, 318*(7177), 162–163.

Rhew, I. C., Hawkins, J. D., & Oesterle, S. (2011). Drug use and risk among youth in different rural contexts. *Health and Place, 17*(3), 775–783.

Rhodes, T. (2009). Risk environments and drug harms: A social science for harm reduction approach. *International Journal of Drug Policy, 20*(3), 193–201.

Rhodes, T., Singer, M., Bourgois, P., Friedman, S. R., & Strathdee, S. A. (2005). The social structural production of HIV risk among injecting drug users. *Social Science and Medicine, 61*(5), 1026–1044.

Sabatier, R. (1996). Migrants and AIDS: Themes of vulnerability and resistance. In M. Haour-Knipe & R. Rector (Eds.), *Crossing borders: Migration, ethnicity and AIDS*. London, UK: Taylor & Francis.

Saggurti, N., Mahapatra, B., Swain, S. N., & Jain, A. K. (2011). Male migration and risky sexual behavior in rural India: Is the place of origin critical for HIV prevention programs? *BMC Public Health, 11*(S6).

Sarkar, K., Bal, B., Mukherjee, R., Niyogi, S. K., Saha, M. K., & Bhattacharya, S. K. (2005). Epidemiology of HIV infection among brothel-based sex workers in Kolkata, India. *Journal of Health, Population and Nutrition, 23*(3), 231–235.

Save the Children. (2012). *Report on Mid-term survey on expanding HIV/AIDS prevention in Bangladesh, RCC program, funded by the Global Fund*. Dhaka, Bangladesh: The Nielsen Company (Bangladesh) Limited.

Singh, N. S. C., Sivek, C., Wagener, M., Hong Nguyen, M., & Yu, V. L. (1996). Determinants of compliance with antiretroviral therapy in patients with human immunodeficiency virus: Prospective assessment with implications for enhancing compliance. *AIDS Care, 8*(3), 261–269.

Smallman-Raynor, M. R., Cliff, A., & Hagget, P. (1992). *London international atlas of AIDS*. Oxford, UK: Blackwell.

Smith, C. J. (2005). Social geography of sexually transmitted diseases in China: Exploring the role of migration and urbanization. *Asia Pacific Viewpoint, 46*(1), 65–80.

Smyth, F. (2005). Medical geography: Therapeutic places, spaces and networks. *Progress in Human Geography, 29*(4), 488–495.

Tempalski, B., & McQuie, H. (2009). Drugscapes and the role of place and space in injection drug use-related HIV risk environments. *International Journal of Drug Policy, 20*(1), 4–13.

Tuan, Y.-F. (1991). A view of geography. *Geographical Review, 8*(1), 99–107.

UNAIDS and IOM. (2001). *Population mobility and AIDS*. Geneva, Switzerland: Joint United Nations Programme on HIV/AIDS (UNAIDS) and International Organization for Migration (IOM).

Vearey, J., Palmary, I., Thomas, L., Nunez, L., & Drimie, S. (2010). Urban health in Johannesburg: The importance of place in understanding intra-urban inequalities in a context of migration and HIV. *Health and Place, 16*(4), 694–702.

Walker, R. (2017). Selling sex, mothering and 'keeping well' in the city: Reflecting on the everyday experiences of cross-border migrant women who sell sex in Johannesburg. *Urban Forum, 28*, 59–73.

Wallman, S. (2001). Global threats, local options, personal risk: Dimensions of migrant sex work in Europe, Health. *Risk and Society, 3*(1), 75–87.

Werb, D., Kerr, T., Fast, D., Qi, J., Montaner, J. S. G., & Wood, E. (2010). Drug-related risks among street youth in two neighborhoods in a Canadian setting. *Health and Place, 16*(5), 1061–1067.

White, N. F. (1981). Modem health concepts. In F. White (Ed.), *N* (pp. 5–18). Toronto, ON: The health conundrum.

Wiersma, E. C. (2008). The experiences of place: Veterans with dementia making meaning of their environments. *Health & Place, 14*(4), 779–794.

Wiles, J. (2005). Conceptualising the importance of place in the care of older people: The role of geographical gerontology. *International Journal of Older People Nursing, 14*(8b), 100–108.

Wiles, J. L., Allen, R. E. S., Palmer, A. J., Hayman, K. J., Keeling, S., & Kerse, N. (2009). Older people and their social spaces: A study of well-being and attachment to place in Aotearoa New Zealand. *Social Science and Medicine, 68*(4), 664–671.

Williams, A. (2002). Changing geographies of care: Employing the concept of therapeutic landscapes as a framework in examining home space. *Social Science and Medicine, 55*(1), 141–154.

Wilton, R. D. (1996). Diminished worlds? The geography of everyday life with HIV/AIDS. *Health & Place, 2*(2), 69–83.

Wilton, T. (1994). Feminism and the erotics of health promotion. In L. Doyal, J. Naidoo, & T. Wilton (Eds.), *AIDS: Setting a feminist agenda*. London: Taylor & Francis.

Wolffers, I., Fernandez, I., Verghis, S., & Vink, M. (2002). Sexual behaviour and vulnerability of migrant workers for HIV infection. *Culture, Health and Sexuality, 4*(4), 459–473.

Wood, E., Chan, K., Montaner, J. S., Schechter, M. T., Tyndall, M., O'Shaughnessy, M. V., et al. (2000). The end of the line: Has rapid transit contributed to the spatial diffusion of HIV in one of Canada's largest metropolitan areas? *Social Science and Medicine, 51*(5), 741–748.

Wood, E., Tyndall, M. W., Spittal, P. M., Li, K., Hogg, R. S., O'Shaughnessy, M. V., et al. (2002). Needle exchange and difficulty with needle access during an ongoing HIV epidemic. *International Journal of Drug Policy, 13*(2), 95–102.

Wood, E., Yip, B., Gataric, N., Montaner, J. S. G., O'Shaughnessy, M. V., Schechter, M. T., et al. (2000). Determinants of geographic mobility among participants in a population-based HIV/AIDS drug treatment program. *Health & Place, 6*(1), 33–40.

Yang, X. (2006). Temporary migration and HIV risk behaviors in China. *Environment and Planning A, 38*(8), 1527–1543.

Chapter 6
Stigmatized HIV Policy Issues and Local Practice

6.1 Introduction

Health policy can be broadly defined as a significant area of government action and interest in health. 'Health policy' is used to encompass any policy which includes strategies and actions undertaken with the aim of maintaining or improving health and providing for the care, treatment or cure of ill-health (Curtis & Taket, 1996). Polices are usually surrounded by conflicts over who is in control and how policies should be implemented. Thus, the issue of health policy is concerned with the political and administrative dimensions involved in health and health care (Blakemore & Symonds, 1997). Over the last two decades, HIV and AIDS have been framed as a 'global problem' (Marx, Halcli, & Barnett, 2012). Hunter (2010) combines anthropology, geography and political economy to chart an interdisciplinary analysis of the uneven geographies of health. As HIV is one of the most concerning issues globally not only for health, but also for social and economic policies, many countries have addressed these issues in different formats with a view to the prevention of HIV.

Many authors highlight the importance of formulating local HIV prevention policies and strategies that acknowledge local models of risk behaviour (for example, Cuadros & Abu-Raddad, 2014). Campbell, Cornish, and Skovdal (2012) discuss the various global scales of analysis of the HIV/AIDS response. In the discussion of sense of 'gender', many studies evidenced that disempowerment is a root source of vulnerability to HIV (Aveling, 2012; Cornish, Campbell, Shukla, & Banerji, 2012; Smith, 2012). They discuss the promotion of gender equality or empowerment of marginalized communities, especially sex workers, is a key pillar of global HIV/AIDS policies. A few studies discuss about the accessibility of appropriate health care services to marginalized or stigmatized people like drug users or HIV diagnosis to PLWH (Bourke, Humphreys, Wakerman, & Taylor, 2012; Collins et al., 2016; Oppong, Tiwari, Ruckthongsook, Huddleston, & Arbona, 2012; Ransome, Kawachi, Braunstein, & Nash, 2016) as intervention strategies against

A. Paul, *HIV/AIDS in Bangladesh*, Global Perspectives on Health Geography,
https://doi.org/10.1007/978-3-030-57650-9_6

HIV infections to reduce the social and economic burdens of HIV/AIDS. Marx et al. (2012) focus on the development of advocacy networks as important actors. In the context of South Asia, Asthana (1996) evaluated Indian policy and legislation related to AIDS, and particularly emphasized 'community participation' (Asthana & Oostvogels, 1996). Khan and Hyder (2001) in a review of HIV/AIDS policy in Pakistan stressed the need for a proactive, organized and integrated policy and programmes that address 'high-risk' behaviours and consider Pakistan's particular social and cultural framework. Effective HIV prevention requires strategies and policies that help reduce the vulnerability of so-called risk groups, like sex workers, to HIV infection by creating a social, legal and economic environment in which prevention is possible.

In Bangladesh, the government adopted a comprehensive national policy for combating HIV/AIDS in 1997 (Panos, 2006). The second National Strategic Plan for HIV/AIDS 2004–2010 was adopted and the third National Strategic Plan was developed by NASP in 2011 guided by a number of strategies and guidelines. The government, in collaboration with NGOs and development partners, has been working in supporting various prevention, treatment, care and support activities among drug users, sex workers, transgenders, PLWH, etc. (NASP, 2014). The fourth National Strategic Plan for HIV and AIDS response was approved by the government in December 2016 for the period 2018–2022 (NASP, 2016). As Bangladesh is a low-prevalence country, its HIV/AIDS responses are prioritized towards prevention, to limit the spread and impact of HIV in the country. In this chapter, I discuss issues related to HIV prevention including addressing marginalization and stigmatization issues that I found important in people in civil society at the grassroots level and among key personnel who plan policy.

6.2 Ambiguity of Recognition to Marginalized People

Sex workers encounter various forms of violence and harassment globally (Lorway et al., 2018). Their activity is not considered as legal in many countries but not illegal either (Fassi, 2011) due to their status and experiences of illegality (Gutierrez-Garza, 2013). Patterson (2015) mentions about de-stigmatization of sex work but Bungay, Halpin, Atchison, and Johnston (2011) emphasize for decriminalization, a contemporary debate issue. Lorway et al. (2018) illustrate the joint role of agency and structure in shaping the everyday risk to sex workers, and resistance (Marshall, 2017). They emphasized 'social environment', particularly the struggle for health protection, safety and financial matters in the everyday life stories of sex workers (see also, Lorraine van Blerk, 2016; Blithe, Wolfe, & Mohr, 2019). Some studies mention a few interventions on behalf of sex workers, for example, economic empowerment (Kuhanen, 2010; Luginaah, 2008), the removal of ambiguities and complexities to their identity crisis (Walker, 2017) and more. They focused on interventions considering the local environment of safety or security giving priority to local knowledge, local tactics and capacities in the intervention approach for both physical and financial risk reduction. Practical difficulties of gaining access to vulnerable groups in Bangladesh, such as commercial sex workers and drug users, have been compounded by a legislative context that reinforces their stigmatization and marginalization (Baden, Green, Goetz, & Guhathakurta, 1994; Kabir, 1989; Khosla, 2009). A NASP (2016) report mentions that there remains a critical programmatic gap within many national responses to HIV with regard to the existence of HIV-related stigma and discrimination. In the policy context, some ambiguities in the recognition of marginalized people are discussed in the following.

'Contradictory Programmes' in HIV Projects In Bangladesh, the sale and use of opiates is strictly prohibited but the state nevertheless provides free syringes through NGOs in the interests of preventing the spread of HIV, at the same time as it harasses addicts. There seems to be a problem in acknowledging the existence of drug users, mitigating their problems and providing accommodation, particularly for the marginalized. There is no legal framework for promoting a 'safe mode' of drug use and the same is true of programmes for brothels, sex workers and the condom issue. There are some contradictory provisions about prostituting girls between government law and the city police actions. The suppression of Immoral Traffic Act, 1933, which states that engaging a woman for immoral purposes against her will is an offence, has some loopholes which not only allow the existence of prostitution but also indirectly help traffickers and pimps to operate their businesses. While in the constitution this type of work is seen as an 'immoral act', there is nevertheless a formal 'affidavit system' which in effect indirectly legalizes the brothel system because it recognizes new girls who are entering a brothel. Also, sex workers have won a verdict from the High Court directing the government to rehabilitate sex workers evicted from brothels. The government have worked for different HIV prevention projects, of which three were focused on sex workers. Although the legal aspect of the profession is rather unclear, all of these above issues indirectly recognize its existence and establishment. A UN official told me that ambiguities in the system cause frustration for the HIV programme:

> 'If you look at the sex workers, are they legal or illegal? The distribution of needles and syringes, is it legal or illegal? We are all working in this limbo between the legal and the illegal! They keep themselves away from the society. That's the biggest frustration in any HIV programme. We need to look at the issue from a non medical perspective'.

In this regard, a government official acknowledges the ambiguity and told me that the government machinery is still not ready to deal with sex workers and drug users. As there are some contradictory laws about the sex workers and drug users issue, so the government operate their 'contradictory programmes' through NGOs. These NGOs have for some years been working with vulnerable people and have been providing condoms and needle syringes in order to reduce the risky nature of their behaviour, as a part of HIV prevention. Due to the stigma attached to working with high-risk groups, especially in the context of AIDS, many government officials are happy to pass the responsibility for HIV prevention activities to the non-governmental sector. The government steers clear because of political sensitivities and employs the NGOs 'at arm's length'.

Religious Sentiment and Brothel Eviction Prostitution is much more common than usually acknowledged in Bangladesh. Although the number of sex working women is unknown, many NGOs and experts estimate that 150,000–200,000 women are involved in both brothel and non-brothel settings. The majority of the CSWs are hotel-, home- or street-based. Due to religious pressure, a few years ago the Magura brothel, close to Jessore, like many other brothels, was forcibly closed and many sex workers were therefore spread out across the Jessore and Khulna region as floating girls and workers in commercial hotels. The effects of brothel evictions, violence against sex workers and the stigma-driven harassment are considered to be associated with AIDS and sex work. Regarding the negative impact of brothel evictions, a local civil society member, who is working in the field with brothel sex workers, told me that:

> 'NGOs give support to the brothel girls, but that the situation is changing. Recently brothels have been closing due to commercial pressure upon the buildings they occupy. High profile physical evictions of the past are no longer happening, fear of human rights violations, but invisible evictions are going on. Within the last few years, two lane near to Jessore brothel have gone, and in the next few years I expect further change. One study shows that 5–6 members of a family are dependent on a brothel girl. If she is evicted then she will not leave the profession but, for her survival, she will either work from home or in the hotel or in the street. If that happens we will lose contact but in the brothel we can monitor her everyday'.

A local UN official also mentioned that some brothels might move from their present locations to government *khas* land (unoccupied land) because of overcrowding in the city centre, as happened with the Daulatdia brothel. But he cannot get the local authorities to discuss it because this is such a 'sensitive issue'. In Bangladesh, decision-making on 'sensitive' issues always has to take religious sentiment into account. For example, any mention of HIV or sexual matters on TV, maybe in an advertisement, is guaranteed to provoke the reaction that this is a breach of religious rules or values. As a result, politicians self-censor because they anticipate an electoral backlash if they stray 'off message' in this area. Some Islamic figures think that if the country can be reminded of its Islamic values, the HIV problem will fade away. They suggest that if men's morality can be improved, brothels will soon have to close.

Hindrance in Recognition of Prostitution Sex workers' poor economic condition and marginalized social identity or status in Bangladesh is the cause of the violation of their rights. Their identity crisis as a group persists despite the recognition given to the sex work profession now being greater than in the past. It is surprising how many NGO people working for the welfare of this marginalized group are not themselves believers. I found one NGO manager at the field level who once helped religiously minded politicians with the eviction of Daulatpur brothel, Khulna, even though he had been working for the NGO with the same sex workers. Dedication is in short supply in this sector of the NGO industry in Bangladesh. Although HIV/

AIDS project workers are aware of the discrimination issue, there is still much prejudice. An NGO high official in Dhaka makes a cautious analysis:

> 'I am confused how long the more positive conception of sex workers can be sustained. As some government projects are going on, so they need to sit with those people, but if the project stops somehow, I don't know whether they will continue to sit with them!'

Although there are some merits and demerits of legalizing prostitution, we should to take the initiative to improve the conditions of the brothel and non-brothel girls because in HIV prevention strategies sex workers are still powerless. But there is still a lack of coordination between different government agencies, like the police and the NGO bodies who work for the sex workers' welfare, in work on accessibility to health rights and reducing harassment. To ensure their health rights, first guaranteeing their human rights is an absolute prerequisite. At a minimum, this requires the sex work profession to be legally recognized by the Government.

Sex Workers' Empowerment In Bangladesh, there are many NGOs working for the welfare of commercial sex workers. Their means of access, however, is mostly hierarchical. My observation is that it is the 'madam' or *sardarni* who benefits on all sides. These 'common faces' (the sex workers' leaders) sit with the NGO and Government officials; they help to make decisions on behalf of the sex workers without any consultation with them and they take monetary benefits from the NGOs. As a result, it is uncertain what true level of empowerment is reaching the subordinates or bonded girls. During my field work, I also observed that NGO people, when they do talk to the sex workers themselves, deal with the leaders or *sardarni* type women in the brothel. As a result, the less vocal ones are not getting proper empowerment training. In this regard, some local NGO officials and key personnel told me that they use the *sardarni* as their strategic way to have access to the others and to change their mind set. Hotel and brothel owners, *sardarni* and the pimps who earn their living from the sex workers' labour all need to be pressured into giving the women some economic freedom.

NGOs could work for improvements in the brothel environment alongside their HIV projects. For example, in the *Maruary Mandir* brothel, Jessore, more than 100 girls have only two usable toilets, which are also used by their customers, children and maid servants. As a result, both toilets are in an unhygienic condition. In order to minimize the brothel sex worker's 'structural vulnerability', recently some NGOs are trying to develop their standard of life by improving utilities like water, sewerage, drainage, gas, rubbish collection and hygiene, so that girls can live with dignity. As their situation improves, so the girls' bargaining powers with customers about condoms will increase. In addition, recently some NGOs have introduced projects to safeguard the girls' money. Usually brothel girls face problems when they go to the bank and feel uneasy. The idea is to help them save money for the future and to provide loans. Another part of this is to educate the non-literate sex workers, so that they can protect themselves from exploitation by their so-called lovers. Promoting human rights and addressing social injustice also can contribute to a reduction in the different forms of stigma and discrimination towards women. Otherwise there is a possibility that the HIV/AIDS epidemic may be feminized. Regarding the sex workers' rehabilitation issue, during the field work I realized that the rehabilitation of sex workers, particularly brothel girls, is problematic because, although the NGOs are giving them training for self-employment or other works, they tend to return to their old profession. One issue is the rate of pay in alternative work by comparison with that in sex work, and also the less labour-intensive nature of the job. Regarding hotel and residence sex workers, training in handicraft making would provide alternative income, along with the creation of savings groups through a micro-credit programme. Then they could be helped to form a cooperative and so establish networks of mutual support.

6.3 HIV Prevention and Limitations

Prevention is a dominant theme in medical and public health discourses. Some authors call for interdisciplinary approaches that move beyond traditional behaviour interventions towards social and structural change (Hefferman, 2002; Parker, Easton, & Klein, 2000; Zierler, Krieger, Tang, et al., 2000). A rights-based approach to HIV/AIDS is argued to be essential to anchor interventions that strengthen people's health rights, thereby supporting them to reduce risky behaviour and vulnerability to infection (Farmer, 1999; Gruber & Caffrey, 2005; Tempalski & McQuie, 2009). A few studies (Luginaah, 2008; Nakku-Joloba et al., 2019; Werb et al., 2010) demonstrate the importance of interventions considering neighbourhood to reduce drug-related harms. In developing countries, 'behavioural change' interventions continue to offer the best opportunity of preventing further spread of HIV/AIDS (Mann et al., 1992; Stover, Walker, Garnett, Salomon, et al., 2002) but little is known about which interventions are most effective (Clift, 1998) partly because of its (behavioural interventions) inadequate definition and the barriers to implementation that are rarely explored (Stephenson, 1999). In public health campaigns, 'community education initiatives' and 'participatory approaches' with many stakeholders need to be considered in the process of identifying health risks and raising awareness of HIV risk and in the design and implementation of community-based interventions for the management of social, attitudinal and behaviour change (Gruber & Caffrey, 2005). In Bangladesh, preventive efforts need to focus on both behavioural and bio-medical risk factors (Islam & Conigrave, 2008; Khosla, 2009; Silverman, Decker, Kapur, et al., 2007). Condom distribution, needle and syringe distribution, behaviour change campaign, STI management and a few harm reduction issues have been emphasized in the strategic plan (NASP, 2016). Some of the following examples show the required HIV prevention approaches for Bangladesh through grassroots voices.

Drug Users' Social Integration The government of Bangladesh has embarked upon both the supply intervention method and the demand reduction method. In India, by comparison, there are a number of different approaches and optional services such as needle exchange programmes, oral drug substitution and treatment programmes. Detoxification is just an interim stage to stabilize the addict, help them to get back their self-confidence and remove their physical and mental dependence on drugs. It is not considered a sufficient intervention in its own right. The idea is that their lives can be changed if they get treatment but the problem is that relapse rates around the world are more than 80–90% after treatment. Long-term treatment and rehab are also needed, along with detoxification. There are many people in civil society who think that NGOs should help recovering addicts by rehabilitating them. Society would benefit if NGOs spent some money on the treatment and rehabilitation of addicts alongside their awareness programmes. Drug use is not only a principal contributing factor to the rise in organized crime, petty thievery and street crime, but also creates social unrest and destabilizes society in Bangladesh. Presently, NGOs are providing them with knowledge but not with the means of applying it in practical terms. Vocational training would be an option because unemployment is a big issue for recovery addicts (RA). An NGO high official sees this holistically:

> 'Actually, we have a good infrastructure at the field level: we know the addicts; we know their problems; we have a good contact and interaction with them; they are coming to our Drop-in-Center and using the services. It's an achievement, and we can go further for more success along with them in a comprehensive approach'.

He also thinks that to implement solutions, there is a need for more skills, and improved attitudes, capacity and knowledge. At the same time, funding is poor and it is not possible to address all of the issues in a comprehensive manner. For example, IDUs need social support along with treatment and they also need 'chain support' to come back into 'normal life'. In this case, NGOs need to start a social rehabilitation programme for drug users, along with their families, to prevent relapses. Recovery from drug addiction is a personal journey that cannot be made alone. Social support is critical. The family is one of the most important social structures that can support recovery and rehabilitation process. It is now recognized that drug addiction is a chronic psychiatric disease and a psycho-social problem. There are relapses and recoveries, as for diseases such as eczema, and addicts are rarely completely cured. There may also be deviations from the addict's own normal social values, maybe including criminal behaviour such as theft to raise money for the next fix. In such cases, a counsellor, a doctor or family members can play a positive role. However, one rehabilitation centre official told me we should be encouraging addicts to improve their mind control rather than giving them medicine in the name of detoxification. A local journalist said that

> 'In the treatment centre, they only provide injections and lots of medicine. What is the gain? Have the addicts received any education? Actually, they become robots, having no emotions, no feelings, and no sense. On the other side, in rehab, they receive education about how to manage withdrawal symptoms and cope with society after treatment. We need to train addicts to increase their mental tolerance level. Actually, the process of mind ventilation is very important to prevent an addict's fall into relapse'.

For Bangladesh, considerable progress can be made if primary-care providers become more knowledgeable about substance abuse, screen their patients for physical symptoms and social problems and make referrals to appropriate treatment programmes. As knowledge about aetiology and access to treatment increases, stigma and denial are likely to decrease. It is an urgent challenge to change the public and political notions about the issues of marginalized communities like drug users in the HIV field. Bangladesh needs to reform its health and narcotic laws, so that it can intensify prevention efforts and improve treatment and care. Under current laws, drug users are arrested and there is no chance to give them access to provide needle syringe or treatment. In summary, more knowledge about substance abuse can contribute to winning the numerous small battles that occur during the recovery process.

'Behaviour Change' Approach As an important prevention part of 'behaviour change', condom promotion is still much stigmatized in Bangladesh. Many people there believe that condom promotion encourages illegal or unethical sex and they ignore arguments about the promotion of safer sex. Due to religious and cultural taboos people, especially in schools, social meetings and public gatherings, are shy of talking about the risks of unprotected sex. Leaders in Bangladesh are wary about speaking directly on HIV/AIDS and other sexually transmitted diseases. Much research around the world shows that condom education, condom skills and condom accessibility help to make for safer sex. But in Bangladesh condoms have become stigmatized because of the HIV awareness campaign. According to key personnel of an international organization, this is because '*now the use of condoms means 'going to prostitutes' and they have negative connotations*'. The simplest and most effective prevention measures, like condom promotion, were not adopted on a large scale. For most Bangladeshis, condoms are known as a means of contraception but are not widely used. Bangladesh government policy emphasizes female-based family planning methods rather than male responsibility through condoms, and all of the health and family planning centres are female focused. So, the female contraception use rate is high but the condom use rate is low. Although all surveillance reports show that HIV awareness is rising among the risk groups, young people and general population, in practice that awareness is not being utilized. In this regard, a UN official told me that male involvement is actually more important than the empowerment of women. This is because the society is male-dominated, so until there is full male awareness about HIV, the part played by women will be less effective.

Regarding the limitations in condom distribution, some NGOs distribute condoms among vulnerable people, particularly sex workers. The idea is that people need to get accustomed to them and then later they may buy them for safe sex. Condoms remain the main method of prevention, though as yet there is no systematic programming strategy for their use. I spoke to NGO officials who claimed that people at risk of contracting HIV, like sex workers, are nowadays more aware about the condom issue than previously due to awareness campaigns and condom distribution programmes. Yet there remains a gap about condom use. One hindrance is that many sex workers do not use condoms in order to please their customers because they know that, if they try to enforce safe sex, there are many poor and abandoned women willing to take their place. At the moment we do not know how many condoms are used in brothels and how many go into the dustbin. One NGO worker told me secretly that many girls throw out the condoms they are given. There is also some confusion between the NGOs that give them away free and those that sell them at a subsidized price. Better planned and integrated distribution would help.

Regarding the condom use dilemma and the increasing number of STD patients, some key personnel, including a reproductive health expert think that increased condom distribution does not mean increased condom use. The test would be if the incidence of STIs had reduced but this is not the case. Due to the sex workers' existing socio-economic conditions, it is difficult to ensure consistent condom use unless clients are motivated. The research suggests that a suitable HIV intervention strategy needs to be developed, considering the socio-economic and cultural aspects, with a provision for continuous monitoring and evaluation. Some key personnel want to follow the 'Thai model' where the aim is 100% condom use but it is the 'Sonagachi model' where social reform movement is preferred, that still predominates. Other than in the relatively ordered environment of brothels, condoms are simply not a priority for the girls working in hotels, residences and in the street, because they cannot earn sufficient to eat properly and are regularly subjected to violence by their customers and harassment by the police. Poverty and chaotic lifestyles make these women at the same time highly vulnerable and hard to reach with health messages.

In response to high STDs among the workers, many researchers in Bangladesh emphasized the implementation of prevention programmes involving effective treatment of STDs, condom and sexual health promotion for CSWs as an emergency basis to check the HIV epidemic in future. Data on the epidemiology of STIs among high-risk behaviour groups is limited in Bangladesh. Availability of diagnostic facilities, poor recognition of STIs as a major public health problem, lack of coordination between service providers and the research community, and poor attendance of STI patients at public clinics and academic institutes are some of the main reasons for lack of STI data. An important further point is that there is a shortage of female doctors in government hospitals and health centres. Female STD patients prefer not to talk about their problems with male doctors due to the stigma of sexual disease. There is a similar problem with the STD-based NGO offices and health centres, which are stigmatized in the eyes of the public. NGO-based STD treatment would be a good tool if it could manage the referral system with the government health centres.

Regarding drug users' risk mitigation as a part of harm reduction approach, it is emphasized safe sex education along with safer injection practices in HIV prevention programmes for injecting drug users. But the various NGO needle syringe exchange programmes have been accused of breaking the law and NGO employees of promoting drugs. Despite such opposition, surveillance reports indicate safer drug use and a decrease in needle sharing. Nevertheless, some of my key informants told me that awareness campaigns alone cannot change risky behaviour, and that, in their opinion, target-wise intervention is essential. In other words, an addict's continuous interaction with peer educators and sourcing of new syringes and condoms can drip feed knowledge about levels of risk. This is risk personalization. In addition, they argued that more advocacy programmes are needed for different stakeholders, including a redirection of police time away from the harassment of addicts towards the disruption of drug smuggling rings. However, public health campaigns target men in the trucking industry in order to increase use in contexts of casual and, in particular, commercial sexual encounters.

6.4 HIV Prevalence, Testing and Awareness

There is debate about the role of male circumcision in the low HIV rate across the world. Although two trials in Kenya and Uganda proved that circumcision can reduce the risk of contacting HIV (Bailey, Moses, et al., 2007; Gray, Kigozi, et al., 2007), some scientists do not agree about its role. In Bangladesh, some researchers have suggested potential reasons for this low HIV prevalence. Sarkar, Islam, Durandin, et al. (1998) emphasize the late introduction of the virus in Bangladesh, Nessa and her colleagues (Nessa et al., 2005) think that, despite high levels of STIs among females with high-risk behaviour, low HIV prevalence might be the result of circumcision. Gibney et al. (1999) say that Bangladesh remains relatively a more insular society than a country like India which has large numbers of tourists and external migrants. The HIV testing and prevalence among the general population is very low in Bangladesh (GOB, 2016). Regarding the awareness programme, in many developing countries, health information is not equally accessible to the less educated, economically disadvantaged and socially marginalized people. Some studies (Gagnon, Merry, Bocking, Rosenberg, & Oxman-Martinez, 2010; Madise et al., 2012; Marshall et al., 2009; Peltzer, Matseke, Mzolo, & Majaja, 2009; Poudel, Jimba, Poudel-Tandukar, & Wakai, 2007) discussed the necessity for improvement of HIV education and access to HIV counselling and testing in different settings. Cianelli et al. (2013) and Ransome et al. (2016) have given importance about the increasing HIV knowledge and HIV testing as prevention technique. In Bangladesh, access to HIV testing centres is limited not only because of inadequate numbers of testing centres but also because of prevailing stigma and discrimination against these groups (NASP, 2016). These lead them to risky sexual behaviour and reduce their sense of vulnerability. Some local voices as policy matters are in the following.

Low HIV Prevalence 'Mystery' There is a 'mystery' about the low HIV prevalence in Bangladesh, as all of the behavioural and bio-medical risks are prevailing there. Some researchers have suggested potential reasons for this low prevalence. There is debate about the role of male circumcision in the low HIV rate across the world. In Bangladesh, it is sometimes said that this Muslim custom may be a factor, although a virologist, who has been working on HIV for a long time, is dismissive of this explanation:

> 'If you ask, why our HIV prevalence is so low despite of the many risk factors, nobody knows. There are so many factors behind it that we need to do more scientific research. I have done research on brothel girls but didn't find any significant rate of infection. I have come to that the conclusion that there has to be something in the Bengali population which is playing a role in the low HIV prevalence. Male circumcision is only practised among the Muslim population, but if you look at Hindu West Bengal the number of PLWH is also low there in comparison with the rest of India. Maybe there are some anthropological and racial matters that are contributing to the low HIV prevalence'.

In the last ten years, many programmes have been started for HIV/AIDS prevention in Bangladesh. Many NGO officials think that despite the many HIV risk factors, Bangladesh's HIV prevalence rate is increasing only slowly and some consider this as a success. His logic is that in many countries of the world, the HIV infection rate has been slow at first, but then there has been a sudden sharp increase. But in the case of Bangladesh, the prevalence rate has not changed significantly. They also consider that Bangladesh has adequate prevention programmes. Although some doubt the quality of work being undertaken, most remain content about the low infection rate of HIV. An alternative view is that the facilities in Bangladesh are so poor that much disease goes undetected. The HIV testing monitoring system is not well coordinated among GOs, NGOs and private medical centres. At the same time, Bangladeshi society is more closed (conservative in a sense) than its neighbour for message sharing. Indian society is more 'open' and has many HIV detection centres. During my interview with key personnel in Dhaka, I found a further potential explanation. Due to cultural values, people do not like to talk about sex and HIV, which are considered a matter of disgrace. They think that if they discuss it with others, even doctors, they will be blamed, and face social and economic discrimination. One NGO chief executive, who works with PLWH, identified three causes of under reporting of HIV in Bangladesh:

> 'Firstly, there is the stigma and discrimination problem. When their status comes out, PLWHs face discrimination from their family first and then others. Secondly, if anyone is diagnosed, he or she will not be able to access proper medical services from anywhere. Third, due to lack of awareness, many potential cases don't know about their own status. Most of the PLWHs in Bangladesh were identified accidentally, when they go to a medical service delivery place for another problem'.

It can be said that the stigma of HIV is playing an important role in under-reporting among the community at risk of HIV, and among the public generally. Most HIV cases are asymptomatic for a long period of time until the onset of AIDS when they recognize their infection for the first time and undergo HIV testing.

Stigmatized HIV Testing at Field Level According to government statistics, up to 2016 there are 128 HIV testing centres throughout the country, of which 18 are operated in public facilities and 110 are operated in NGO settings. HIV testing or Voluntary Counselling and Testing (VCT) activities were started in Bangladesh in 2006. Few organizations follow the rapid test kid system according to WHO guidelines. As HIV tests have become available mainly in urban-based NGO settings, many of these testing programmes fail to ensure confidentiality and cannot provide access to prevention information or treatment. In addition, the government is failing to address the widespread stigma faced by those testing positive. At the field level, fear, stigma and discrimination are restricting HIV testing. In order to find out about attitudes, I talked with one VCT centre manager in Khulna, who deals with the clients of local prostitutes. He explained the field level problem about HIV testing:

> 'Actually, people feel fear the loss of confidentiality. When we manage awareness-raising meetings, many people show an interest in testing and tell us about their problems and risk behaviours but only a few of them come later to the centre for testing. Most seem to think that if they are tested, they will be identified as HIV-infected person and they will be an even worse situation in their family and society. Counselling about confidentiality can play a positive role. We need to provide more awareness about avoiding social discrimination. Sensitization meetings on the issue in the family and in society have a role in increasing people's mental tolerance levels'.

Although some NGOs have developed testing centres in Dhaka and other big cities, their numbers are insufficient. The lack of trained counsellors is also proving to be a handicap and more training programmes are needed if the number is to be increased. As many people fear testing, the 'de-stigmatization' of other medical service points can play a positive role in this regard. The government could launch an anti-stigma programme concerning at health care centres. Unless there are treatment support facilities near counselling and testing centres, AIDS patients will continue to be left in a vulnerable position. An expert told me that VCT programmes would be more acceptable to the public if the country's medical service centres and providers could be de-stigmatized:

> 'If you find a positive, then what will you do? Where will you send the patient for treatment? That patient will be discriminated against wherever he or she goes, so you should to de-stigmatize the neighbouring clinics or hospitals of VCT centres because that patient would then not have to travel far away for treatment. Up to now this has not happened for the present VCT centres'.

However, some NGO officials doubt the notion of confidentiality in VCT centres. They said that 'if you keep a matter very confidential, it may cause harm', whereas an 'open secret' can have good results. Regarding sensitization to reduce fear and stigma, some NGO officials suggest advocacy workshops for religious people. They think that if the religious leaders can be properly sensitized, this would help to create a congenial atmosphere and ultimately reduce stigma and discrimination. Being an Islamic country, Bangladesh could take advantage of engaging '*imams*' (religious personalities) and Islamic leaders in active prevention programmes.

'Standard Awareness Campaign' HIV/AIDS is sometimes regarded as a problem confined to foreigners and highly marginal groups. Awareness is a big issue for encouraging people to be tested and the gauge a more accurate level of HIV in Bangladesh. But general awareness raising programmes on HIV/AIDS still focus mostly on how HIV is spread, and there is very limited discussion of human rights, discrimination or stigma. Some NGO employees that I spoke to are of the opinion that HIV awareness campaigns as they are presently constituted actually increase stigma about HIV and appear to convince people that a positive diagnosis is a 'death sentence'. Different NGOs are campaigning differently according to their client group and there is no integration. An NGO executive in Dhaka suggested that a 'national standard campaign' for reducing misconceptions would increase social acceptance:

> 'We need a campaign to lessen misconceptions about HIV and increase the social acceptance of positives. Here we are campaigning for the sake of our target group, others are targeting sex workers, and still others for drug users. The campaigning techniques are different, some emphasizing STDs, while others focus on the high mortality of HIV/AIDS. There is no integration among the different groups, and there is a danger of confrontation between us and those who are saying that AIDS means death. They are trying to persuade people to refrain from the risky behaviour, but in reality they are responsible for increasing stigma and reducing the likelihood that at-risk people will go for testing. We need to coordinate our means of campaign implementation'.

There is no doubt that the intention behind these campaigns is awareness-raising. But due to common understanding of the general people it has virtually turned into a 'death sentence' for PLWH. This stigmatized campaign or myths about HIV have been creating a new sect of untouchables. As a result, PLWH in many cases victimize themselves, thinking that their lives have no worth, whereas in reality the symptoms can be controlled. A few NGOs disseminate messages that becoming PLWH does not mean the end of the world. Here, the media has the potential to create mass awareness of HIV/AIDS, to promote positive attitudes towards people living with HIV/AIDS, and to influence people to change the high-risk behaviour that makes them vulnerable to the infection.

6.5 Discrimination, Support and Treatment for PLWH

HIV/AIDS still perceived and experienced as a stigmatized and technically challenging 'disease' (Kielmann et al., 2005), especially to health care practitioners (Bennett, 2005) for their concern with the risk at which they put themselves, and potentially their families (Mason, Carlisle, Watkins, & Whitehead, 2001). People living in rural and remote areas face challenges in accessing appropriate health services (Bourke et al., 2012), especially for ARV (Gilbert & Walker, 2009). Collins et al. (2016) demonstrate the urgent need to consider the sifting of HIV care services. According to the NASP (2016), there are 1,964 registered members of PLWH networks who receive free ART as they fulfill the national eligibility criteria: CD4<350, WHO clinical stage 3 or 4, pregnancy or co-infection with TB. The current ART coverage of HIV patients constitutes a half of the known PLWH (1,964/3,922) and only 20% of the estimated total PLWH (for example, Islam, Rasin, & Rahman, 2013). According to UNAIDS (2015), ensuring human rights of HIV-infected people is an effective weapon in the fight against HIV/AIDS (Ghimire, 2007). Some local policy voices are in the following.

Discrimination and Hypocrisy Prejudice may interfere with confidentiality and consideration of the patients' emotional well-being. In Bangladesh, PLWH face much discrimination, such as the refusal to offer medical support or forcing them to leave hospital. This behaviour comes from all levels of health providers, from doctors and nurses to laboratory technicians. Many practitioners change their attitude as soon as they know the patient's status. In my interviews, this situation prompted NGO officials to ask how long such human rights violations will continue towards PLWH. Since the first cases detected about 30 years ago, hospital staff have received little training in the risks of treating PLWH. As a result, fear and misunderstanding are universal. The problem appears to be one of attitude and unwillingness to deal with positives. At the field level, an international NGO manager at local level told me of his experience in this respect:

> 'Many doctors are not interested to listen to the messages in our advocacy workshops. Actually, they have an arrogant mentality that they know everything. Although doctors know the real cause of HIV and its means of transmission, their attitude is still unhelpful for positives. Maybe deep down they don't trust their own knowledge and don't want to take any risks'.

HIV/AIDS still perceived as a stigmatized and challenging health problem. In this regard, a UN official replied that most of the health sector is not geared to deal with PLWH. Consequently, he thinks that, although the patient has rights to receive treatment, at the same time doctors and other service providers also have rights to protect themselves. I was told that an NGO request for a separate room for PLWH in a local hospital is blocked by physicians who ask 'where is the government policy for a separate room?' A leading NGO official responsible for PLWH was scathing about the negative attitude of physicians:

> 'Doctors' training in theory and in practice is quite different. We ask the government body to send doctors to our centre to deal with positives, but what is happening now? Some doctors go abroad to get training in the clinical management of HIV, so their visiting card changes but not their behaviour towards the positives. Some don't allow any referred positive patients into their consulting rooms and outside they breach the patient's confidentiality and identify them as PLWH. It's hypocrisy and discrimination against PLWH.

The interrelation of HIV/AIDS and human rights is mainly based on the logic that HIV-infected and affected people should also enjoy fundamental and other rights such as the right to health, right to education, right to property and the same right to legal treatment as other people. To eliminate the various forms of discrimination against HIV-infected and affected people and to ensure their fundamental human rights, then there is a great need for the restructuring of existing social and legal settings. There is a further need to assure society that PLWH can also live a dignified and meaningful life if they are respected by society. In the Bangladeshi constitution, everyone has the right to access government health facilities. So, legally speaking, there is no need to adopt any new policy for PLWH alone. Moreover, there should be effective laws to address the violation of human rights of HIV-infected people.

Breast Milk Issue and Dilemma The risk of transmitting HIV through breastfeeding is hard to measure. Most children with HIV infection probably became infected before they were born, or while they were being born. Breast-feeding by PLWH mothers is a sensitive issue for stakeholders and policy planners. Worldwide it is generally recommended that breast-feeding is avoided because of possible infection. But in Bangladesh there is generally a presumption in favour of breast-feeding because the babies of poor mothers would probably otherwise be malnourished. Formula milk is too expensive for them to buy and at present NGOs are not providing this or other alternatives such as cow's or goat's milk. A virologist is very critical of donors' role in this respect:

> 'The donors' logic is that we will not able to bear the cost of ready food but how many PLWH mothers do we have? Can we not provide the money from our programmes to mothers for baby food? There is money for condoms, so why not also for baby food?'

I found a new born boy of a PLWH couple who was given formula milk up to the age of two years to check mother-to-child transmission of the HIV from breast milk and he has been eating normal food for the last six months. Before the birth of the boy, the mother was given ART and nutritional food by one NGO for a year.

Care and Medical Support In Bangladesh, policy initiatives such as care and treatment including ART for people already affected by HIV/AIDS are being implemented on a very limited scale. It is urgent to ensure ART for PLWH who are in need. The current coverage of treatment, care and support to PLWH is low. About one-fifth of the estimated PLWH are receiving ART in proper time due to the present procurement process. According to NASP, the number of locations from which ART is available is currently limited to 10, of which 4 are public hospitals and 6 are NGOs. But there is a need for ART access throughout the country. In addition to managing ART, these centres should be able to deal with a range of issues like palliative care, nutrition support and referral care for both HIV-related and non-HIV-related complications. However, the nutritional support to PLWH needs to be integrated into the existing programme. Regarding cost sharing matters of ARV, most of the other NGO chiefs think that the state should have responsibility for distributing ARV free of charge. In India, ARV is provided free to positives and at the same time they are available in the market. Their policy is that those who can afford it, but do not want to disclose their identity, can buy from the local market. In Bangladesh, the government may in future distribute ARV to positives through NGOs, but there are some questions. First, will they be available in the market? The government will need to give thought to selling the medicine in the open market for those who do not want to disclose their identity. The second question is the debate about whether the idea of free medicine should be extended to other patients, such as those with cancer. One NGO chief commented:

> 'Firstly, ARV is very costly and there are many other issues involved with an HIV case. For example, there is stigma attached to this disease which is absent for other diseases. Secondly, if you have a disease, you will need to take the drugs for a certain time to finish the course, as is also the case with tuberculosis, but in case of HIV you need to take the drug for your whole life, until death'.

Most of the debate concerning the availability and access to antiretroviral therapy (ART) now revolves around the core issues of economics, equity and ethics. A mix of payment-free, subsidized and self-paying systems is applied by governments, and criteria for access to ART differ widely. In this respect, some officials think that the government can arrange for ARV to be free only for a certain period of time. HIV/AIDS patients who temporarily stop taking their antiretroviral drugs to reduce side effects or because of supply disruption are doubling the risk of getting full blown AIDS or are likely to cause resistance to treatment. Second line drugs are the only hope for sustainable life, yet these drugs are much more expensive. At present, some NGOs provide first line treatment for HIV free of cost, but not the second line.

The National Strategic Plan has set a goal of ensuring that anyone infected with HIV/AIDS has access to comprehensive systems of care, support and treatment by the year 2022. In this connection, Bangladesh can follow the Brazilian model which has proven that it is possible to contain HIV/AIDS in a resource-poor environment with a relatively weak health infrastructure. This model ensures the right to free access to treatment for HIV-infected people and others who have opportunistic infections, a strong relationship between the government and civil society groups, including religious people to reduce stigma and discrimination.

6.6 NGO 'Politics' and Donor Policy

The empowerment of marginalized communities and the process of morality increase for decreasing inequality and corruption are one of the main issues in HIV discourse (Cornish et al., 2012; Smith, 2012). A significant proportion of HIV/AIDS prevention activities in Bangladesh are implemented by NGOs (Bhuiya, Rob, Yusuf, & Chowdhury, 2004; NASP, 2016). Currently, the funding for the NASP comes via the government from the different funding bodies including World Bank. Many international NGOs and missionary agencies provide funding directly to NGOs (Paul, 2009). The UN and other partners focus on reducing stigma, discrimination and violence against people living with and affected by HIV through technical support, advocacy, orientation and education (, 2016; NASP, 2014). It has been reported that each of these funding agencies disburses funds according to its own mandate and specific objectives, rather than the needs of the population (Panos, 2006). As a result, the coverage of the ongoing intervention programmes has been reduced in many places which will make difficult to achieve 'Ending AIDS by 2030' in Bangladesh (NASP, 2016). There are some quality questions in HIV projects and some 'unhealthy' competition among NGOs discussed in the following found in the 'ground'.

Quality Questions in HIV Projects Many NGOs are working well with HIV projects in Bangladesh to supplement government activities but there are also a few anomalies. For instance, during my field work, I visited a local NGO in Khulna, where no qualified doctor is available to see STD patients but every week they compile a list of people who have visited their drop-in treatment centre. I heard that although only two patients had been seen by a paramedic, twenty cases were reported. Such exaggeration is presumably to show the donors that they are active with STD patients. In the case of condom distribution, there is also evidence that they are also overstating the figures and taking the opportunity to misappropriate funds by selling condoms in the open market. I also observed several NGOs with staff of too low a quality to implement their projects, and I heard that some employees do not get their proper salary due to 'salary surrender' (a significant proportion of salary they have to capitulate) to some high officials. Some of the drug users themselves complain about outreach workers and NGO doctors that they lack responsibility and efficiency. Despite all of these limitations, limited resources and the many risk factors, there are well-intentioned NGOs that are developing their capacity. An NGO high official thinks that:

> 'Some NGOs don't know the techniques for making scientific size estimations of the problem. Though they have government projects, their knowledge of how to handle that group is limited. As a result, there is a question about the quality of these programmes'.

Regarding the issue of transparency, a UN official told me that there is a need to monitor projects more intensively to expose cases of misappropriation of funds. Several key personnel consider that we need to monitor the government's role, the donor's role and NGOs' role in every development project because accountability has two sides: one concerns the service providers and other the project participants. Accountability is not possible unless the target group understand the project and services which they can expect from the providers. The people have to be empowered first so that they can seek services in an assertive way. If the people are aware, then sustainability is possible. On the other side, donors should be more vigilant and they also need coordination among themselves to reduce the duplication of services. According to an NGO executive:

> 'In India policy planners bargain with the donors but how much can we bargain with the donors? Are we handling matters skilfully? There is presently a lack of patriotism and there are mind-set problems'.

Thus although, there have been some positive achievements from HIV projects in the last ten years including awareness and addition to the school curriculum, many organizations working in this field are not maintaining quality. Although NGO workers have more experience than government workers in the field of HIV prevention, there are still some doubtful practices.

Invisible 'Conflicts' Among NGOs Newly 'discovered' PLWH are subject to a kind of 'competition' among NGOs to visit them. They may go with government officials and local representatives and this creates curiosity among neighbours and other local villagers. This creates a negative impact on their status in their family, in the locality and, more importantly, in the patient's own mind. Visits therefore lack confidentiality and some positives flee the scene; a few commit suicide out of shame and desperation. Despite recent attempts to reduce stigma and discrimination, there is much work to do to change this mentality. Many PLWH and civil society people accused some NGOs that they look at this sensitive matter from a professional perspective rather than on humanitarian grounds. They 'recruit' as many new positives as possible because then they will attract additional funds and receive credit for their work in the HIV field. Some PLWH angrily told me that some, although not all, NGOs are running 'profitable businesses'. In other words, they inflate the number of PLWH in their programmes in order to get more funds. An NGO executive elaborated on this:

> 'In Bangladesh, PLWH are used now as a 'token' by some NGOs. Those NGOs are giving help, showing hope of foreign tours, dreams of leadership. Positives are thinking that this is for their benefit but they do not understand that the benefit is going to the NGO and that in truth the future impact will be bad. This is the way many PLWH are being victimized. The NGO which has no expertise in care and support for HIV is trying to say that they have skilled staff and are able to handle care and treatment projects and they are looking for funds'.

This 'unhealthy' competition among NGOs in the provision of HIV services is happening because there are not enough funds in this area. The government response is poor and there is misuse of foreign funds through the distribution of money among non-compliant organizations through the tendering process. Evidence suggests that a number of Bangladeshi NGOs are motivated by self-promotion rather than altruism. Many organizations would appear to be more interested in attracting international funds and attending international conferences than working in the communities they claim to serve. There seem to be some NGOs which lack appropriate expertise and experience. Sometimes they develop a consortium with existing NGOs in the field and so are able to access projects. In this regard, one NGO chief told me:

> 'Fund distribution is not equal because some funds are not going to the 'right person'. Some NGOs have bid for projects who previously worked in micro-credit. They joined with other NGOs, made a good project proposal and finally got the project. They also maintained a 'good channel' with high officials'.

I observed that there is an invisible conflict and lack of co-operation between the NGOs working at the field level, and this has negative impacts on the stakeholders. For example, they will not comment on the medication regimes of their clients in the name of secrecy. There is overlap in service provision but no integration or coordination of effort. At the moment, the 'confidentiality' of their PLWH stakeholders is one reason given by NGOs for not talking to each other.

PLWHs' Desire and 'Complaints' PLWH seek many kinds of psycho-social, economic and treatment support from the NGOs. The latter try to provide medical support, usually irregular, such as ARV, nutritional support, hospital admission, referral, training and advocacy. But an NGO high official comments on how expectations cannot always be met:

> 'Their demands depend upon their status. Some positives cannot afford basic food at all and others need support to balance their nutrition. Some need money for medicines because

their immunity to disease is low. Some need support to maintain their confidentiality. Others worry about who will take care of their children when they die, or just need help with their babies. Some unmarried PLWH want to get married and have a baby'.

Some NGOs make little impact. In Khulna, a nationally renowned NGO started a nutrition support project among PLWH. After six months, this stopped without explanation as described by one PLWH:

'(That NGO) gave us some support, monetary help, but they stopped the project and in the meantime we had become dependent on this help. Now many of us are suffering. We need vitamins and nutritious food. We utilized the money, but now they have stopped it, so how can we continue the nutritious food?'

One local NGO chief identified the waste in many HIV projects. He commented how some Dhaka-based NGOs hold workshops or awareness raising sessions in the field for a few days and then return to Dhaka.

Donors' Policy Limitations Donors play a vital role in HIV prevention. All of the HIV projects in Bangladesh are fund-based and designed according to the donor's wish. The coordination of donor activities is weak; leads to frequent overlapping of services but more comprehensive packages are needed. In HIV projects in Bangladesh, there are many issues which are obligated by different donors and their conditionality. This seems to me to be a limitation in HIV prevention in Bangladesh. Firstly, there is a place for outreach programmes that combine needle exchanges with treatment. But due to USA policy obligations, NGOs who are taking USAID funds cannot use a harm reduction approach towards drug addicts. They give treatment, rehabilitation and detoxification for addicts but needle exchange is not possible. The US government feels that needle exchange promotes greater addiction, although there is another view that needle exchange programmes are working well in some countries. For the same reason, US-funded NGOs are also prevented from using oral drug substitution and condom promotion. Even the terminology is restricted: advocating sex worker's rights is not possible because US policy does not acknowledge prostitution as a legitimate form of 'work'. Second, although, there is a proverb that beggars cannot be choosers, Bangladeshi NGOs resent the 'conditionality' put upon donor funds and any 'interference', as they see it, with their ongoing projects. But as they depend upon the donors for their sustainability, NGOs need to accept the strings attached, even though they may consider them inconvenient and even unethical.

However, I observed that some international agencies take funds from international donors and then give funds to local NGOs. But, at the same time, the international donor is also distributing money to the same NGOs. So local NGOs have 'multiple relations' with donors and there are sometimes 'cross donations' due to a lack of coordination. For example, in one small area, I found five NGOs providing almost the same services with the help of different donors' funds. Third, due to contractual mechanisms, inefficiency and donor level bureaucracy concerning the release of money, NGOs frequently find that they are in between projects with no funds to manage the transition. Meanwhile HIV transmission continues. There are politics here at the donor levels that are partly the result of the culture of large organizations. Such donors are powerful and impose their policy on NGOs, with the result that decision-making at the local level is not possible. Some donors lack flexibility. One NGO programme manager told me that during the programme, they could not make any long-term plans due to the uncertainty of the project and its extension, and this hampered the project's achievements. He thinks that if they could know the total duration earlier, they would be able to make plans for the long term.

It is mentionable that World Bank, various bilateral agencies and global fund have major contribution in response to HIV/AIDS programme since 2000 along with the support from government of Bangladesh. For example, HIV/AIDS Prevention Project (HAPP) was the first major projects (2004–2007) under NASP which was supported by World Bank and DFID. HIV/AIDS Targeted Intervention (HATI) was supported by the World Bank (2008–2009) financed Health, Nutrition and Population Sector Programme (HNPSP). HIV/AIDS Intervention Services (HAIS) programme was supported by World Bank (2009–2011) financed HNPSP. The HIV/AIDS Prevention Services (HAPS) programme was supported by the HPNSDP (2011–2016). Most of the projects focused on intervention packages for different high-risk groups including sex workers, drug users, PLWH and general people and implemented through mainly non-government collaborations. Moreover, advocacy, capacity building and programme support were the main components of these projects. HPNSDP continues to allocate funds (2017–2022) to prevention among key populations and international migrants and care, treatment and support for PLWH with an aim to minimize the spread of HIV on the individual, family and community level.

Due to donor dependence, NGOs cannot work consistently or continuously. But with projects lasting only a year or two, they cannot prevent HIV, and so sustainability is a problem. NGO projects that have been initiated with outside assistance may not prove sustainable once that assistance has been withdrawn. In order to overcome the uncertainty, some NGOs try to establish 'core funds' for their future. An official of International organization said '*We need to develop a contingency plan and need to make a widow fund so that we can continue the project during the crisis time or bridging period*'. Recently, a global policy in HIV prevention has come to a critical point. In HIV prevention projects, the ABC model (Abstinence, Be faithful to your partner, use a Condom) has been successfully used in many places in the world. Medical research has repeatedly and clearly shown that it is better access to condoms and greater use of these lifesavers in safer sex, which has the greatest impact on reducing the risks of dying from AIDS. In the 16th International AIDS conference in Toronto, Canada, Microsoft founder Bill Gates, as a keynote speaker said that the ABC approach had saved many lives but with some limitations. He emphasized the social changes necessary to stop the spread of HIV/AIDS, for example, the empowering of women. However, these are very broad social changes needed that will eventually end stigma and sexual prejudice and giving women the laws, social rights and medical care to protect themselves.

6.7 HIV Surveillance and Strategic Planning

There is a necessity for appropriate structural interventions for high-risk areas (Feldacker, Emch, & Ennett, 2010; Goldenberg et al., 2011) along with important strategic planning to address HIV risk. First, surveillance systems are used to estimate HIV/AIDS prevalence and incidence and for identifying subgroups of the population at high risk (Matsuyama, Hashimoto, et al., 1999) using some sensitivity (Khaw, Salama, Burkholder, & Dondero, 2000; Khawaja, Gibney, Ahmed, & Vermund, 1997). In Bangladesh, AIDS control mechanisms are not well integrated with the basic public health care infrastructure facilities including surveillance system (Khan, 2005; Panos, 2006; Rahman, 2005). A nationwide serological surveillance was last done in 2011 and a comprehensive behavioural surveillance has not been done since 2006–2007 (NASP, 2016). Second, although blood screening can reduce the risk through transfusion, HIV is still transmitted through transfusions (Khaw et al., 2000). In Bangladesh, blood is not always screened for HIV, and much of the supply is from professional blood donors (Akhter et al., 2016; Quader, 2004). No data is available from private hospitals and other non-government actors regarding HIV screening of donated blood (NASP, 2016). Third, the global commercial sex industry promotes trafficking and prostitution (Williams, 1999). This sector remains relatively under focused and there is no strong measure to curb trafficking (Paul, Atkins, & Dunn, 2012). Fourth, adolescents in Bangladesh receive inadequate information about HIV and reproductive health since discussions about sex, drugs and HIV are taboo and the flow of information is obstructed by religious beliefs and cultural traditions (Khan, 2002; Reeuwijk & Nahar, 2013). Finally, although demand for condom use is gradually increasing among vulnerable populations, supplies of condoms remain irregular (Panos, 2006). Some recommendations are discussed in the following using the grassroots voices.

HIV Surveillance In Bangladesh, HIV/AIDS surveillance has always been accorded low priority in national planning and resource allocation. This has resulted in inappropriate epidemiological data, causing confusion in policy planning and policy failure. For behavioural surveillance, there is limited coverage but sero-surveillance is more widespread. Programmes of sero-surveillance involve approaching organizations for help in managing sentinel sites and sampling sites. In that case, neutral venue may reduce the error. ICDDR,B has been conducting surveillance for HIV in the country on behalf of the Bangladesh government in collaboration with other partners. Surveillance data have been used to monitor the progress of the HIV epidemic and changes in risk behaviour over time. These data have also been used effectively in mobilizing and directing resources appropriately. Though, since 2007, there was no comprehensive behavioural surveillance and after 2011, no sero-surveillance was conducted in Bangladesh, more interaction between NGOs is needed to make the data more representative and therefore reliable for strategic planning purposes. These discussions should involve people at the grassroots for site and group selection. Moreover, a major constraint identified in formulating evidence-based policies is the limited availability of local data on HIV/

AIDS. It is important to conduct nationwide serological and behavioural surveillance on a regular basis for future directions. Therefore, source of funding for regular serological and behavioural surveillance should be ensured. Many key officials raise another issue about the 'clearance' order of surveillance reports or inconsistencies in the reporting from different government ministries and AIDS-related committees. In this regard, a high official of international organization commented on how reporting delays and documentation by the government are the main hindrances in translating research findings into action. For surveillance, technical committee members select potentially vulnerable groups for sampling on the basis of available information and resource constraints. Resource limitations are the main obstacle to sampling more sites and people. The border regions are seen by some to be on the front line of risk behaviour and vulnerability for HIV transmission but historically HIV surveillance on the border with India, particularly port areas, was not included intensively. In other words, the importance of geographic place has been largely overlooked in terms of surveillance. Following my analysis in the previous chapters and naturalistic observation, many HIV risk elements are concentrated in those areas and there is a need for place awareness to be included in surveillance strategies as a matter of priority. An HIV expert commented:

> 'A sudden HIV epidemic will start in our border areas which may gradually develop. Nobody will know the real situation unless we include the border areas in the surveillance properly'.

However, urban areas are the main foci of risk, much more so than rural areas, and the slums are the most vulnerable places of all. Creative thinking about interventions might include trying to reach those who live abroad with messages about minimizing their risk behaviour. Highly mobile workers, such as construction and transport workers, who have little AIDS prevention knowledge and are away from their families, are more likely to engage in high-risk behaviours.

Strategic Planning There is a need for HIV-related research projects for understanding the epidemic and for improving the effectiveness of the response. One local NGO official remarked on how the theory of programme organization is very different to practice. He wants a review of HIV prevention strategies in order to manage resource utilization properly. A UN official thinks that Bangladesh needs to give priority to surveillance, capacity building and blood safety rather than to HIV mainstreaming (including HIV components in every development issue). And an international organization personnel considers that making strategic plans based upon surveillance reports is all very well but in many cases these become 'wish lists' of ambitious planning. He prefers working 'to the point' of what can really be achieved. In Bangladesh, various strategies, guidelines, management plans, training manuals, toolkit and national consultation plan have been developed by funding bodies, GO and NGOs in last ten years to reduce specially stigma and discrimination related to HIV/AIDS among the marginalized community. But no remarkable achievement or visible progress is yet to mention. The country lacks a proactive mechanism for HIV/AIDS-related project monitoring and evaluation for quality control. However, there are some issues which need to be considered in the future planning for the sake of greater interest in HIV prevention, as discussed below.

Condom Supplies Although demand for condom use is gradually increasing among vulnerable populations, the lack of condom availability from time to time is a visible marker of the inefficiency of the system. There seem to be problems with estimation procedures and the implementation of planning. The building of a buffer stock is an obvious starting point and then a better set of procurement procedures. Otherwise stock outs undermine the success of programmes, leading to unintended pregnancies and the spread of infection. As the Social Marketing Company (SMC), the government and UNFPA are dealing with condom issue, so they need more coordination to prevent shortages.

Female Condoms My key informants had varying views about female condoms. Some believe them to be an important tool to protect girls who cannot protect themselves and have no negotiation skills. Until these women are able to insist upon condom use by their sex partners, the female condom is an effective alternative for safer sex, especially in the case of alcohol and drug users. On the other hand, some think that the female condom is not well accepted for a number of reasons. It is not practical, either for the sex worker or for the client. Clients who do not like the male condom also tend to reject the female condom because they can feel it during intercourse. Only if they are drunk or semi-conscious will they not be aware of it. Also, it is costly, so there are commercial considerations, and there are associated risks because, if the girls do not wash it properly, there is a heightened chance of STD cross-infection from one client to another.

Reproductive Health Adolescents receive inadequate information about HIV and reproductive health in Bangladesh due to many social and religious taboos. Specially, female adolescents in Bangladesh are not sufficiently aware of AIDS. It is recommended that strong efforts are necessary to improve awareness and to clarify misconceptions about AIDS through improving access to education, mass media and condom use promotion. Regarding the need for reproductive health education, one NGO programme manager told me that in the long term it will be useful. Although the government has recently taken the initiative to include a chapter on HIV in a textbook, many NGO staff feel that this is inadequate, requiring more illustrative discussions and training for teachers to overcome issues of shyness about sexual and reproductive issues.

Women Trafficking Trafficking in human beings, especially women and children, is a crime that violates all sorts of human rights and dignity. Bangladesh has been reported as one of the top countries having the highest incidences of trafficked women. They usually end up in brothels in Kolkata or Mumbai and are highly vulnerable to HIV/AIDS. Therefore, Bangladeshi policy planners should consider this issue not only from a health perspective but also as a matter of development.

Blood Transfusions Although blood screening has reduced the health risk through transfusion in most parts of the world, HIV is still transmitted through transfusions. In Bangladesh, professional blood donors play a significant role and most hospitals in Bangladesh lack blood screening facilities. Apart from these government centres, there are many commercial blood banks operate illegally. Almost no rules are followed by the private blood banks for collection, testing, processing, storage and distribution of blood. The government needs to implement initiatives to improve blood testing infrastructure and blood products with quality control, promote voluntary blood donation, develop and improve facilities for plasma fractionation and improve the management, monitoring and evaluation of blood transfusion services. There is also a need for better control over the licensing and inspection of private blood banks.

6.8 Expected Role of Government in HIV Prevention

Government is the most important stakeholder in fighting HIV/AIDS, but the various political parties and politicians could take up the HIV/AIDS agenda as an important priority. Few studies (Seckinelgin, 2012; Stover et al., 2002) show the impact of the governance of success in HIV/AIDS policy and necessity of political commitment in HIV prevention for effective use and mobilization of funds and programme implementation. There have been calls for Bangladesh to mobilize political will to act rapidly and decisively to avert an HIV epidemic before it becomes too late (Mahmood, 2007). In Bangladesh, many NGOs, AIDS service organizations and civil society organizations have been implementing programmes/projects in different parts of the country (Hossain, 2007). NASP is responsible for facilitating overall coordination of the national response to HIV/AIDS in Bangladesh. But presently, NASP is not structured with sufficient capacity to promote overall programme planning, coordination, monitoring and evaluation, system wide information sharing and single programme steering (NASP, 2016). Monitoring and evaluation systems would seem to be essential for the purposes of learning, understanding and improving programme performance and impact (NASP, 2014). Some policy issues are critically discussed below.

Political Commitment and State Role Government and society generally must also acknowledge the scale of HIV/AIDS and take the initiative to reduce the associated stigma. At the moment, this is almost absent in the political arena of Bangladesh where there is a reluctance to speak about the issue. It has been proven that governance, either good or bad, has a direct effect on the HIV/AIDS epidemic. Instances of good governance marked by sincere political commitments to combat HIV/AIDS in Thailand, Uganda and Senegal have shown positive results in stemming its spread with the cooperation of the donor nations. In Bangladesh, many NGOs, AIDS service organizations and civil society organizations have been implementing programmes/projects in different parts of the country. These initiatives have focused on prevention of sexual transmission of the virus among high-risk groups. There are some self-help groups working with PLWH patients, providing counselling and financial help to them. But, despite funds from different global initiatives, the Bangladeshi government's response to the management of AIDS patients has been weak. Regarding their expectations of government, almost all PLWH argue that politicians need to take greater responsibility for their care and treatment. They criticize government expenditure on seminars rather than on welfare. One PLWH commented that:

> 'On world AIDS day the government always gives assurances. Ministers seek to reassure us but in truth they depend upon the NGOs. In seminars they make promises, but implement only a few works. They give responsibility to the NGOs and we are forced to survive with some help from these organizations. We have a health department and, given the political will, the government could provide us with treatment. We are not demanding centres in every district, but the government could establish a centre in Dhaka with every facility'.

Patients also think that the government should emphasize HIV prevention in a more extensive way. In doing so, they should use PLWH people to raise awareness. They believe that if government could utilize PLWH in the struggle for HIV prevention, it may be more effective than any present work. However, advocacy issues are considered to be a priority activity to create awareness and mobilize both political commitment and relevant ministries and departments for the national response on HIV and AIDS.

NASP Role for Coordination Under the ministry of health and family welfare, NASP is responsible for facilitating overall coordination of the national response to HIV/AIDS in Bangladesh. Its role is mainly coordination, monitoring, policy strategy formulation and guidance to all stakeholders in the HIV field. It has a stewardship role and a mandate to coordinate the activities of all organizations for HIV prevention. But presently, NASP is not structured with sufficient capacity to promote overall programme planning, coordination, monitoring and evaluation. The weak leadership of the NASP hinders the effective implementation of programmes at ground level. My field work uncovered many criticisms of the NASP. One district deputy civil surgeon, for instance, openly stated that there are GO/NGO coordination problems: NGOs are not aware of government activities and vice versa. Many NGOs are implementing targeted and other interventions in different geographical areas in an uncoordinated and overlapping manner without system wide prioritization, coordination or information sharing. One local elected municipal chairman told me that '*the government has started a programme through the NGOs but they do not monitor it or follow it up. The government needs to play a more active role*'. A NASP executive told me on the success and limitations of the NASP:

> 'At least we can say we are trying to contain the epidemic with involvement from international organizations. With the support of these partners, we have many achievements. At the same time, we have many constraints, many resource limitations, and strong religious values, so many things remain to be done. It is true that we could not cover all of the most at-risk populations through the various packages because the funds were not sufficient and the initial size estimation was not properly done'.

Internal collaboration and coordination are needed to avoid problems of duplication. This should be dealt with at the heads of programmes or donor level. Ideally resource allocation would be modified on the basis of real need. One suggestion is for one NGO to be responsible for STI treatment, another for peer education work and so on. Regarding proper fund utilization, a key person of an NGO suggested that the government develop education materials for distribution among NGOs by generating a central department for communication materials development. At the same time, the government could provide training for NGOs on the basis of proper guidelines. These kinds of central activities for HIV projects would save money. However, sidelining this intrinsic care for PLWH, NASP has been spending time, resources and energy in organizing conferences and seminars which are unreachable to the majority of HIV/AIDS patients. One local NGO chief thinks that in Bangladesh, there is no GO-NGO coordination and actually the NGO people do not even like each other and do not want to see each other being successful. It can be said that in order to have optimum coordination in HIV projects, Bangladesh needs to develop an effective coordination mechanism in the NASP and NAC.

Limitations of NASP The weak leadership hinders the effective implementation of programmes at field level. For example, many key personnel are critical about the NASP's unstable human resource policy. If an official stays for a few years, s/he would gain valuable experience by being trained, making foreign visits and being exposed to the HIV field. But frequent government transfers of personnel disrupt this capacity building. According to a UN official:

> 'Among our limitations, we have a shortage of resource persons and technical persons and we wish that we had more trained the people in the field who really understand the epidemic, are motivated to work, and would really like to see the epidemic come to an end. Often I have felt that it's a problem that people are very much motivated but they don't have the skills to use the millions of dollars coming into our country. I suppose quite obviously that's the case, but nonetheless we find it very difficult when we want to do something. Sometimes it takes 15 days to move a paper from one table to another, which is really sad. We are famous for bureaucracy, so we always start our all work 3–4 months behind'.

In addition, Bangladesh has no national body for monitoring and evaluation of its HIV projects, though they are essential for learning, understanding and improving programmes. It is also necessary for accountability for financial and programme stewardship. In order to strengthen monitoring, a national independent body is needed which will monitor all HIV work on a regular basis. This would employ experts and work on behalf of government or under state control. Funds for this work could be top-sliced from the total HIV project allocations. A high official of international organization commented that '*If you look at the programme critically, you can find many gaps. But some high officials may forbid you to be too critical because we are not interested to share our weaknesses*'. Moreover, there are some irregularities and a lack of transparency about NGO programmes, so some people raise questions about their capacity.

6.9 Concluding Remarks

Substantial sums have come to Bangladesh in the last 20 years for different HIV projects and this is a good time to ask questions about the optimum utilization of these funds. These should include the matter of poor coordination and why it has not been possible to develop referral systems between organizations. Networks are presently not working properly, and duplication of effort is partly due to the lack of coordination and strategic thinking. It is partly also due to the idea that the broader the base of service provision the greater will be the power to attract funds. It is very difficult to overcome this mentality. Since the high officials in the projects are carrying out the donors' wishes, it is at the donor stage that the strategic thinking needs to take place. For instance, contingency plans need to be developed in order to sustain projects in between funding periods. So, instead of thinking of innovative methods, unfortunately it is necessary for the time being to think about the quality and coordination of existing works. HIV projects would be more fruitful if government could choose the right people, the right organization and the right activities to do the work. Otherwise there will duplication, under-reporting or even over-reporting and the epidemic will not be checked. So, the government must handle programmes more carefully in terms of quality and coverage. The good NGOs who are working well should be identified, along with those that need a capacity build-up to do their work properly.

References

Akhter, S., Anwar, I., Akter, R., Kumkum, F. A., Nisha, M. K., Ashraf, F., et al. (2016). Barriers to timely and safe blood transfusion for PPH patients: Evidence from a qualitative study in Dhaka, Bangladesh. *PLoS One, 11*(12), e0167399.

Asthana, S. (1996). AIDS-related policies, legislation and programme implementation in India. *Health Policy and Planning, 11*(2), 184–197.

Asthana, S., & Oostvogels, R. (1996). Community participation in HIV prevention: Problems and prospects for community-based strategies among female sex workers in Madras. *Social Science and Medicine, 43*(2), 133–148.

Aveling, E.-L. (2012). Making sense of 'gender': From global HIV/AIDS strategy to the local Cambodian ground. *Health and Place, 18*(3), 461–467.

Baden, S., Green, C., Goetz, Anne-Marie and Guhathakurta, M. (1994). *Background report on gender issues in Bangladesh, Prepared for the British high commission, Dhaka* (BRIDGE Report no. 26). Sussex, UK: Institute of Development Studies, University of Sussex.

Bailey, R. C., Moses, S., Parker, C. B., Agot, K., Maclean, I., Krieger, J. N., et al. (2007). Male circumcision for HIV prevention in young men in Kisumu, Kenya: a randomised controlled trial. *Lancet, 369*, 643–656.

Bennett, A. (2005). *Culture and everyday life*. London, UK: Sage.

Bhuiya, I., Rob, U., Yusuf, N., & Chowdhury, A. H. (2004). *Bangladesh database on HIV/AIDS: South Asia political advocacy project*. Geneva, Switzerland: Population Council and UNAIDS.

Blakemore, K., & Symonds, A. (1997). Recent trends in health policy: Consumerism and managerialism. In S. Taylor & D. Field (Eds.), *Sociology of health and health care* (2nd ed.). Tokyo, Japan: Blackwell Science.

Blithe, S. J., Wolfe, A. W., & Mohr, B. (2019). *Sex and Stigma: Stories of everyday life in Nevada's legal brothels*. New York, NY: New York University Press.

Bourke, L., Humphreys, J. S., Wakerman, J., & Taylor, J. (2012). Understanding rural and remote health: A framework for analysis in Australia. *Health and Place, 18*(3), 496–503.

Bungay, V., Halpin, M., Atchison, C., & Johnston, C. (2011). Structure and agency: Reflections from an exploratory study of Vancouver indoor sex workers. *Culture, Health & Sexuality, 13*(1), 15–29.

Campbell, C., Cornish, F., & Skovdal, M. (2012). Local pain, global prescriptions? Using scale to analyse the globalization of the HIV/AIDS response. *Health and Place, 18*(3), 447–452.

Cianelli, R., Villegas, N., Lawson, S., Ferrer, L., Kaelber, L., Peragallo, N., et al. (2013). Unique factors that place older Hispanic women at risk for HIV: Intimate partner violence, machismo, and marianismo. *Journal of the Association of Nurses in AIDS Care, 24*(4), 341–354.

Clift, E. (1998). IEC interventions for health: A 20 year retrospective on dichotomies and directions. *Journal of Health Communication, 3*, 367–375.

Collins, A. B., Parashar, S., Closson, K., Turje, R. B., Strike, C., & McNeil, R. (2016). Navigating identity, territorial stigma, and HIV care services in Vancouver, Canada: A qualitative study. *Health and Place, 40*(July), 169–177.

Cornish, F., Campbell, C., Shukla, A., & Banerji, R. (2012). From brothel to boardroom: Prospects for community leadership of HIV interventions in the context of global funding practices. *Health and Place, 18*(3), 468–474.

Cuadros, D. F., & Abu-Raddad, L. J. (2014). Spatial variability in HIV prevalence declines in several countries in sub-Saharan Africa. *Health and Place, 28*(July), 45–49.

Curtis, S., & Taket, A. (1996). *Health and societies: Changing perspectives*. London, UK: Arnold.

Farmer, P. (1999). *Infections and inequalities: The modern plagues*. Berkeley, CA: University of California Press.

Fassi, M. N. (2011). Dealing with the Margins of Law: Adult sex workers' resistance in everyday life. *Oñati Socio-Legal Series, 1*(1), 36.

Feldacker, C., Emch, M., & Ennett, S. (2010). The who and where of HIV in rural Malawi: Exploring the effects of person and place on individual HIV status. *Health and Place, 16*(5), 996–1006.

Gagnon, A. J., Merry, L., Bocking, J., Rosenberg, E., & Oxman-Martinez, J. (2010). South Asian migrant women and HIV/STIs: Knowledge, attitudes and practices and the role of sexual power. *Health and Place, 16*(1), 10–15.

Ghimire, R. (2007) HIV/AIDS and human rights. Editorial, *The Rising Nepal*, January

Gibney, L., Choudhury, P., Khawaja, Z., Sarker, M., Islam, N., & Vermund, S. H. (1999). HIV/AIDS in Bangladesh: An assessment of biomedical risk factors for transmission. *International Journal of STD & AIDS, 10*, 338–346.

Gilbert, L., & Walker, L. (2009). "They (ARVs) are my life, without them I'm nothing"-experiences of patients attending a HIV/AIDS clinic in Johannesburg, South Africa. *Health and Place, 15*(4), 1123–1129.

GOB. (2016). *DHIS2, DGMIS, DGHS, Ministry of Health and Family Welfare*. Dhaka, Bangladesh: Government of Bangladesh.

Goldenberg, S. M., Strathdee, S. A., Gallardo, M., Nguyen, L., Lozada, R., Semple, S. J., et al. (2011). How important are venue-based HIV risks among male clients of female sex workers? A mixed methods analysis of the risk environment in nightlife venues in Tijuana, Mexico. *Health and Place, 17*(3), 748–756.

Gray, R. H., Kigozi, G., et al. (2007). Male circumcision for HIV prevention in men in Rakai, Uganda: A randomised trail. *Lancet, 369*, 657–666.

Gruber, J., & Caffrey, M. (2005). HIV/AIDS and community conflict in Nigeria: Implications and challenges. *Social Science and Medicine, 60*(6), 1209–1218.

Gutierrez-Garza, A. (2013). *The everyday moralities of migrant women: Life and labour of Latin American domestic and sex workers in London*. PhD thesis, The London School of Economics and Political Science (LSE), London, UK.

Hefferman, C. (2002). HIV, sexually transmitted infections and social inequalities: When the transmission is more social than sexual. *International Journal of Sociology and Social Policy, 22*(4), 159–176.

Hossain, M. R. (2007). AIDS: Epidemic bell ringing, *The Daily Star*, October 28, Dhaka.

Hunter, M. (2010). Beyond the male-migrant: South Africa's long history of health geography and the contemporary AIDS pandemic. *Health and Place, 16*(1), 25–33.

Islam, M. M., & Conigrave, K. M. (2008). HIV and sexual risk behaviors among recognized high-risk groups in Bangladesh: Need for a comprehensive prevention program. *International Journal of Infectious Diseases, 12*(4), 363–370.

Islam, M. S., Rasin, S., & Rahman, L. (2013). Antiretroviral Therapy (ART) programme in Bangladesh: Increasing national ownership. *Sexually Transmitted Infections, 89*(Suppl 1), A378–A378.

Kabir, M. A. (1989). *Comparative study of institutionalized prostitutes in Tanbazar and floating prostitutes in Dhaka City*. Programme and information workshop on development strategies for prevention and control of STDs in Bangladesh, Dhaka.

Khan, M. A. (2002). Knowledge on AIDS among female adolescents in Bangladesh: Evidence from the Bangladesh demographic and health survey data. *Journal of Health Population and Nutrition, 20*(2), 130–137.

Khan, M. A. (2005) AIDS crisis looming: A disaster waiting to happen? *The Daily Star*, April 10, Dhaka.

Khan, O. A., & Hyder, A. A. (2001). Responses to an emerging threat: HIV/AIDS policy in Pakistan. *Health Policy and Planning, 16*(2), 214–218.

Khaw, A. J., Salama, P., Burkholder, B., & Dondero, T. J. (2000). HIV risk and prevention in emergency-affected populations: A review. *Disasters, 24*(3), 181–197.

Khawaja, Z. A., Gibney, L., Ahmed, A. J., & Vermund, S. H. (1997). HIV/AIDS and its risk factors in Pakistan. *Editorial review, AIDS, 11*, 843–848.

Khosla, N. (2009). HIV/AIDS interventions in Bangladesh: What can application of a social exclusion framework tell us? *Journal of health, population and nutrition, 27*(4), 587–597.

Kielmann, K., Deshmukh, D., Deshpande, S., Datye, V., Porter, J., & Rangan, S. (2005). Managing uncertainty around HIV/AIDS in an urban setting: Private medical providers and their patients in Pune, India. *Social Science and Medicine, 61*(7), 1540–1550.

Kuhanen, J. (2010). Sexualised space, sexual networking & the emergence of AIDS in Rakai, Uganda. *Health and Place, 16*(2), 226–235.

Lorway, R., Lazarus, L., Chevrier, C., Khan, S., Musyoki, H. K., Mathenge, J., et al. (2018). Ecologies of security: On the everyday security tactics of female sex workers in Nairobi, Kenya. *Global Public Health, 13*(12), 1767–1780.

Luginaah, I. (2008). Local gin (akpeteshie) and HIV/AIDS in the Upper West Region of Ghana: The need for preventive health policy. *Health and Place, 14*(4), 806–816.

Madise, N. J., Ziraba, A. K., Inungu, J., Khamadi, S. A., Ezeh, A., Zulu, E. M., et al. (2012). Are slum dwellers at heightened risk of HIV infection than other urban residents? Evidence from population-based HIV prevalence surveys in Kenya. *Health and Place, 18*(5), 1144–1152.

Mahmood, S. A. I. (2007) Leadership is a must to combat this scourge. *The Daily Star*, December 1, Dhaka.

Mann, J. M., et al. (Eds.). (1992). *AIDS in the world: The global AIDS policy coalition*. Boston, MA: Harvard University Press.

Marshall, A.-M. (2017). *Confronting sexual harassment: The law and politics of everyday life*. London, UK: Routledge.

Marshall, B. D. L., Kerr, T., Shoveller, J. A., Patterson, T. L., Buxton, J. A., & Wood, E. (2009). Homelessness and unstable housing associated with an increased risk of HIV and STI transmission among street-involved youth. *Health and Place, 15*(3), 783–790.

Marx, C., Halcli, A., & Barnett, C. (2012). Locating the global governance of HIV and AIDS: Exploring the geographies of transnational advocacy networks. *Health and Place, 18*(3), 490–495.

Mason, T., Carlisle, C., Watkins, C., & Whitehead, E. (Eds.). (2001). *Stigma and social exclusion in healthcare*. London, UK: Routledge.
Matsuyama, Y., Hashimoto, S., et al. (1999). Trends in HIV and AIDS based on HIV/AIDS surveillance in Japan. *International Journal of Epidemiology, 28*, 1149–1155.
Nakku-Joloba, E., Pisarski, E. E., Wyatt, M. A., Muwonge, T. R., Asiimwe, S., Celum, C. L., et al. (2019). Beyond HIV prevention: Everyday life priorities and demand for PrEP among Ugandan HIV serodiscordant couples. *Journal of The International AIDS Society, 22*(1), e25225.
NASP. (2014). *Revised 3rd national strategic plan for HIV and AIDS response 2011–2017, National AIDS/STD Program*. Dhaka, Bangladesh: Ministry of Health and Family Welfare, Government of Bangladesh.
NASP. (2016). *Fourth national strategic plan for HIV and AIDS response, National AIDS/STD program*. Dakha, Bangladesh: Ministry of Health and Family Welfare, Government of Bangladesh.
Nessa, K., Waris, S. A., Alam, A., Huq, M., Nahar, S., et al. (2005). Sexually transmitted infections among brothel-based sex workers in Bangladesh: High prevalence of asymptomatic infection. *Sexually Transmitted Diseases, 32*(1), 13–19.
Oppong, R., Tiwari, C., Ruckthongsook, W., Huddleston, J., & Arbona, S. (2012). Mapping late testers for HIV in Texas. *Health and Place, 18*(3), 568–575.
Panos. (2006). *Keeping the promise? A study of progress made in implementing the UNGASS declaration of commitment on HIV/AIDS in Bangladesh*. Dhaka, Bangladesh: The Panos global AIDS Programme.
Parker, R. G., Easton, D., & Klein, C. H. (2000). Structural barriers and facilitators in HIV prevention: A review of international research. *AIDS, 14*(Suppl. 1), S22–S32.
Patterson, C. B. (2015). Beyond the stigma: The Asian sex worker as First World savior. In M. Laing, K. Pilcher, & N. Smith (Eds.), *Queer sex work* (pp. 53–65). London, UK: Routledge.
Paul, A. (2009) *Geographies of HIV/AIDS in Bangladesh: Vulnerability, stigma and place*. Durham theses, Durham University. http://etheses.dur.ac.uk/1348/
Paul, A., Atkins, P. J., & Dunn, C. E. (2012). Borders and HIV risk: A qualitative investigation in Bangladesh. *Oriental Geographer, 53*(1), 73–82.
Peltzer, K., Matseke, G., Mzolo, T., & Majaja, M. (2009). Determinants of knowledge of HIV status in South Africa: Results from a population-based HIV survey. *BMC Public Health, 9*(174).
Poudel, K. C., Jimba, M., Poudel-Tandukar, K., & Wakai, S. (2007). Reaching hard-to-reach migrants by letters: An HIV/AIDS awareness programme in Nepal. *Health and Place, 13*(1), 173–178.
Quader, G. M. (2004). HIV and AIDS: Present situation and future challenges. *The Daily Star*, November 29, Dhaka.
Rahman, A. Z. M. (2005). HIV/AIDS: Bangladesh country situation, World AIDS Day Special. *The Daily Star*, December 1, Dhaka.
Ransome, Y., Kawachi, I., Braunstein, S., & Nash, D. (2016). Structural inequalities drive late HIV diagnosis: The role of black racial concentration, income inequality, socioeconomic deprivation, and HIV testing. *Health and Place, 42*(November), 148–158.
Reeuwijk, M. V., & Nahar, P. (2013). The importance of a positive approach to sexuality in sexual health programmes for unmarried adolescents in Bangladesh. *Reproductive Health Matters, 21*(41), 69–77.
Sarkar, S., Islam, N., Durandin, F., et al. (1998). Low HIV and high STD among commercial sex workers in a brothel in Bangladesh: Scope for prevention of larger epidemic. *International Journal of STD & AIDS, 9*, 45–47.
Seckinelgin, H. (2012). The global governance of success in HIV/AIDS policy: Emergency action, everyday lives and Sen's capabilities. *Health and Place, 18*(3), 453–460.
Silverman, J. G., Decker, M. R., Kapur, N. A., et al. (2007). Violence against wives, sexual risk and sexually transmitted infection among Bangladeshi men. *Sexually Transmitted Infections, 83*, 211–215.
Smith, D. J. (2012). AIDS NGOS and corruption in Nigeria. *Health and Place, 18*(3), 475–480.

Stephenson, J. M. (1999). Evaluation of behavioural interventions in HIV/STI prevention. *Sexually Transmitted Infections, 75*(1), 69–71.

Stover, J., Walker, N., Garnett, G. P., Salomon, J. A., et al. (2002). Can we reverse the HIV/AIDS pandemic with an expanded response? *The Lancet, 360*, 73–77.

Tempalski, B., & McQuie, H. (2009). Drugscapes and the role of place and space in injection drug use-related HIV risk environments. *International Journal of Drug Policy, 20*(1), 4–13.

UNAIDS. (2015). *Report on the global AIDS epidemic*. Geneva, Switzerland: Joint United Nations Programme on HIV/AIDS (UNAIDS).

van Blerk, L. (2016). Livelihoods as relational Im/mobilities: Exploring the everyday practices of young female sex workers in Ethiopia. *Annals of the American Association of Geographers, 106*(2), 413–421.

Walker, R. (2017). Selling sex, mothering and 'keeping well' in the city: Reflecting on the everyday experiences of cross-border migrant women who sell sex in Johannesburg. *Urban Forum, 28*, 59–73.

Werb, D., Kerr, T., Fast, D., Qi, J., Montaner, J. S. G., & Wood, E. (2010). Drug-related risks among street youth in two neighborhoods in a Canadian setting. *Health and Place, 16*(5), 1061–1067.

Williams, B. (1999). *Working with victims of crime: Policies, politics and practice*. London, UK: Jessica Kingsley publishers.

Zierler, S., Krieger, N., Tang, Y., et al. (2000). Economic deprivation and AIDS incidence in Massachusetts. *American Journal of Public Health, 90*(7), 1064–1073.

Chapter 7
Summary and Conclusion

7.1 Introduction

HIV/AIDS is no longer a problem just for individuals or nations: it has turned into a global issue. Every World AIDS day (1st December), people across the globe stand together as an opportunity of solidarity against HIV/AIDS because there is still no vaccine to prevent the disease and no medicine to cure those who are already infected by it. It has become one of the most harmful diseases humankind has ever faced. But in many countries, policy makers initially perceived HIV to be just another public health problem rather than seeing it as a socio-economic threat or a national emergency. In Bangladesh, although the HIV/AIDS epidemic is not well researched, HIV is at least being detected among the 'most at-risk' groups. There are many risk behaviours for HIV transmission, such as homosexuality, polygamy and illicit drug use, which are discouraged according to Muslim values but which still occur in practice. In the following sections, a summary of the geographies of HIV/AIDS in Bangladesh is discussed from the context of stigmatized people, place and policy.

7.2 HIV/AIDS and Challenges for Bangladesh

In Bangladeshi society, several contextual features, including widespread poverty; the often subordinate status of women, including their role as marginalized sex workers; unequal access to health services; and low literacy and education are responsible for enhancing health risks. In addition, there is a lack of multi-sectoral integration, long-term strategy, policy level response and commitment for empowerment of vulnerable groups to negotiate issues like stigma and discrimination. Moreover, operational research, which is essential for the effective implementation

A. Paul, *HIV/AIDS in Bangladesh*, Global Perspectives on Health Geography,
https://doi.org/10.1007/978-3-030-57650-9_7

of any health programme, is limited. During fieldwork, I found different symbolic comparisons and concepts relating to the risks and vulnerabilities of HIV/AIDS. Some civil society participants commented that 'AIDS is spreading fast like electricity in Bangladesh' or 'Bangladesh is on an HIV bomb which will burst at any time'. These expressions serve to symbolize the concerns of civil society and to demonstrate the challenge of containing HIV/AIDS in a setting where conservative values inhibit people from discussing sex and related diseases, even with physicians. The stigmatized notion of 'sexualization' of the HIV epidemic has prevented people, including health officials, from considering non-sexual routes of transmission. The stigma attached to HIV infection is related to under-reporting, and restricted testing and counselling facilities. Low levels of knowledge and heightened fears among medical service providers towards PLWH are also exacerbating infection risk and bring about a lack of preparedness. Although there is an extensive awareness programme for reducing HIV risk behaviours, concerns remain that people are not employing preventative measures in practice.

From a geographical point of view, stigmatized places and high mobility are commonly cited as reasons for the rise of HIV/AIDS in Bangladesh. Frequent movement of the population into neighbouring countries where HIV rates are high may be an important source of entry of cases. Apart from women trafficking, injecting drug use is also more prevalent in border areas. Areas bordering India such as Manipur have particularly high rates of HIV infection among drug injectors. This research has uncovered how vulnerable groups, notably sex workers, drug users and transport workers are subject to heightened health risks due to the places they frequent. Rural-urban migration, border crossing for livelihoods, women trafficking and the out-migration of unskilled people deserve further research in order to understand the 'channelling' of risk from one place to another.

Marginalization of Sex Workers Sex workers in Bangladesh face exploitation and social ostracism. Prostitution is a largely invisible aspect of Bangladeshi society, but it is a readily available component of everyday life, particularly in urban areas. The growing level of cautiousness about HIV among clients and sex workers is affecting incomes while income uncertainty itself, in turn, brings raised health risks. Harassment increases the vulnerability of sex workers to sexually transmitted diseases and they are often victims of violence. Most aim to secure payment from clients before services are rendered and safe sexual behaviour is prioritized where possible. But all groups of sex workers reported violence: rape, robbery and beatings by both police and *mastaans*. Female sex workers in Bangladesh are commonly identified as a potential 'reservoir of infection' because of their limited choices. Men's refusal to use condoms or to stop relations with other partners denies women opportunities to protect themselves. In addition, most brothel and non-brothel sex workers find it difficult to maintain proper hygienic practices after attending a customer not only due to a lack of knowledge concerning its necessity but also because of the absence of hygienic facilities. This lack of knowledge and practice therefore serves to fuel potential health risks.

Importantly, the stigmatized identity of sex workers means exclusion from mainstream society. Many street-based, hotel or residence-based sex workers suffer from depression and many attempt or consider suicide due to the high social stigma, economic vulnerability and physical violence. Stigma associated with identity and health risks may therefore negatively affect their 'lifeworlds'. Violence, low self-esteem and the dominant role of men impact on the vulnerability of these marginalized women to HIV infection. Table 7.1 summarizes the different kinds of problems which commercial sex workers face.

Given that NGOs provide condoms and run awareness programmes for brothel-based sex workers, this group may be categorized as having lower health risk but their social status as perceived by others, including their family and relatives, is very low. This low social status (stigmatized recognition) and poor living standards and condition cause frustrations among this group which may also indirectly impact on their health risk. In hotel- and residence-based sex workers fear for losing 'good reputation' from neighbours, other family members, etc. is very high as they always try to hide their present profession. Hotel managers' or residence madams' exploitation ensures financially uncertainty that can exacerbate health risks for these groups. Street-based girls may be categorized as having very low living standards and diminished social status. Their health risks are also raised as their voices are unheard in issues such as requests for condom use by clients. Floating girls also have to face severe health risks as they are compelled to prioritize financial income over place of transaction or condom use.

Vulnerability of Drug Users Narratives of drug users who took part in this research serve to highlight the physical vulnerability and risks they face every day. Addictive behaviour poses different challenges to their lives. Different addict subgroups develop strong interpersonal bonds, similar images of appropriate and acceptable narcotic behaviour, a language and customs of their own and feelings of being rejected by society. Addicts' 'lifeworlds' centre on drug taking with little time given over to hygiene-related practices such as taking a bath, washing clothes, brushing teeth and social activities such as communicating with family members. Addictive behaviour is shaped by individual factors and by social problems such as financial crises or homelessness. Most drug users' poor health is related to these everyday uncertainties. Withdrawal symptoms and physical suffering, money earning for feeding the drug habit, stealing and needle sharing all are interlinked between drug use and its users. In addition, many addicts are considered as at-risk for HIV

Table 7.1 A comparison of different sex workers' lives

Type	Living standard	Social status	Identity fear	Health risk
CSW—Street	Very low	Worst	Medium	Worst
CSW—Floating	Low	Low	High	Worst
CSW—Residence	Medium	Suspicious	Very high	Medium
CSW—Hotel	Low	Suspicious	Very high	Worst
CSW—Brothel	Low	Very low	High	Low

transmission because of their unprotected sexual behaviours, blood selling and low awareness of HIV/AIDS.

During my field work, it became apparent how opiate drugs, particularly heroin, are products for which addicts have never-ending desires. My interviewees told me that when an addict earns 50 taka, s/he consumes heroin worth the whole amount and if s/he could earn 200 taka s/he would also consume the whole amount. Addicts consider heroin to be their main 'food' and they do not feel hungry for other foods such as rice before or after taking heroin. In Bangladeshi society, people are more like to despise the addict rather than the addiction while this marginalization exacerbates the addict's need for drugs. Figure 7.1 illustrates this cycle of vulnerability in terms of economic marginalization and social stigma.

Dejected PLWH Discrimination and prejudice are forms of social exclusion responsible for separating individuals from society. HIV has the most profound impact on PLWHs' lives, as most face discrimination which creates 'negative' dimensions of their identity. In Bangladesh, a lack of economic power reflects the marginalized status of most people with HIV/AIDS. Medical and social aspects of HIV/AIDS are interconnected with the degree of stigma and psychological complications that are experienced by the PLWH person. As stigma is the outcome of

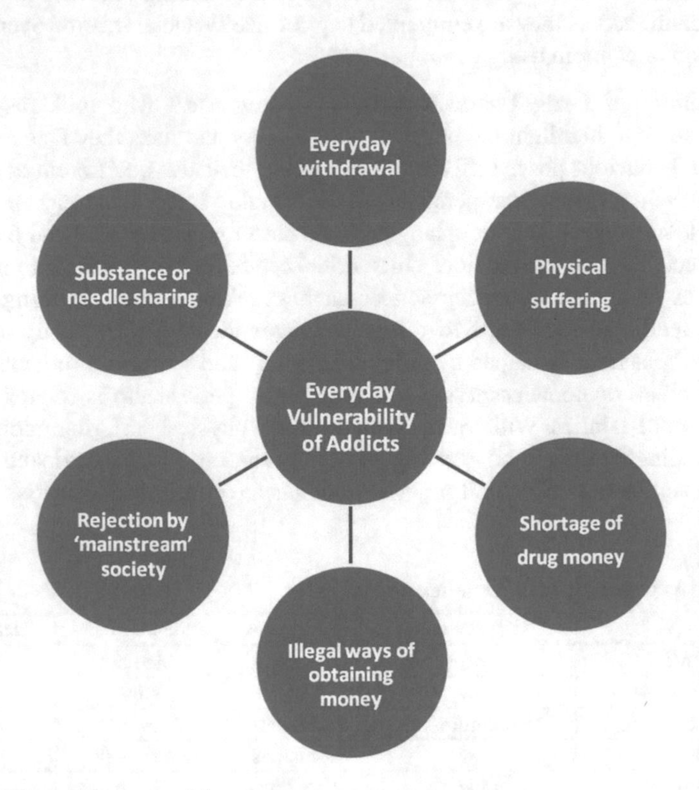

Fig. 7.1 Vulnerability cycle of drug users

widespread fears and perceptions of risk, PLWHs have to tolerate aspects of discrimination and neglect every day. Three interrelated dimensions of the illness are important: physical impacts with opportunistic infection, psychological changes and social responses with negative public reactions. Most of the stigma bearers experience a sense of shame and tend to employ strategies to overcome this and try to present themselves as 'normal'. Most PLWH people in Bangladesh experience a range of social and emotional difficulties including separation from families and friends, loss of key roles, disruption of plans for the future, loss of self-image and self-esteem, and uncertain and unpredictable futures.

Stigma and psychological impacts change over time, depending on the patient's physical condition and the real and perceived efficiency of the patient's medical treatment. From a social perspective, women with HIV are more likely to be victims of domestic violence and to lack financial support. They often experience fear, isolation and uncertainty, heightened by the likelihood of abandonment. Discrimination against PLWH has been growing but successive governments have been accused of not adequately addressing the issue of social stigma—the biggest factor in the way of PLWH try to lead normal lives while combating their condition. There is a need for legislative reforms addressing such wide-ranging issues as discrimination, ethics, access to treatment, privacy and confidentiality. Current legislation offers a very little support to HIV sufferers. In addition, the role of religious institutions in combating HIV in the south Asian region could be more proactive. From the present research, it is clear that discrimination against PLWH is very common in Bangladesh. Discrimination and negative identities are inter-related and self-reinforcing, with raised health risks as a by-product. Figure 7.2 shows the links between discrimination and raised health risks including the impact of a stigmatized identity on treatment-seeking behaviour.

7.3 Is HIV/AIDS Only a Health Concern?

The HIV/AIDS pandemic has shown a consistent pattern through which marginalization, discrimination, stigmatization and, more generally, a lack of respect for human rights and dignity of individuals and groups heighten people's vulnerability to becoming exposed to HIV. In the present research, individual biographies provide clear evidence of everyday life constituting a number of social, cultural and economic challenges for key marginalized groups in terms of HIV risk. In Bangladesh, HIV/AIDS is still, however, considered to be only a health problem or the concern of the individual. Socio-economic and human rights perspectives are notably missing. In many lower-income countries, HIV/AIDS is now considered a threat for overall economic development and to be connected with aspects of globalization and therefore a multidisciplinary issue. But in south Asia, social reaction to marginalized people who are vulnerable to HIV further fuels the crisis.

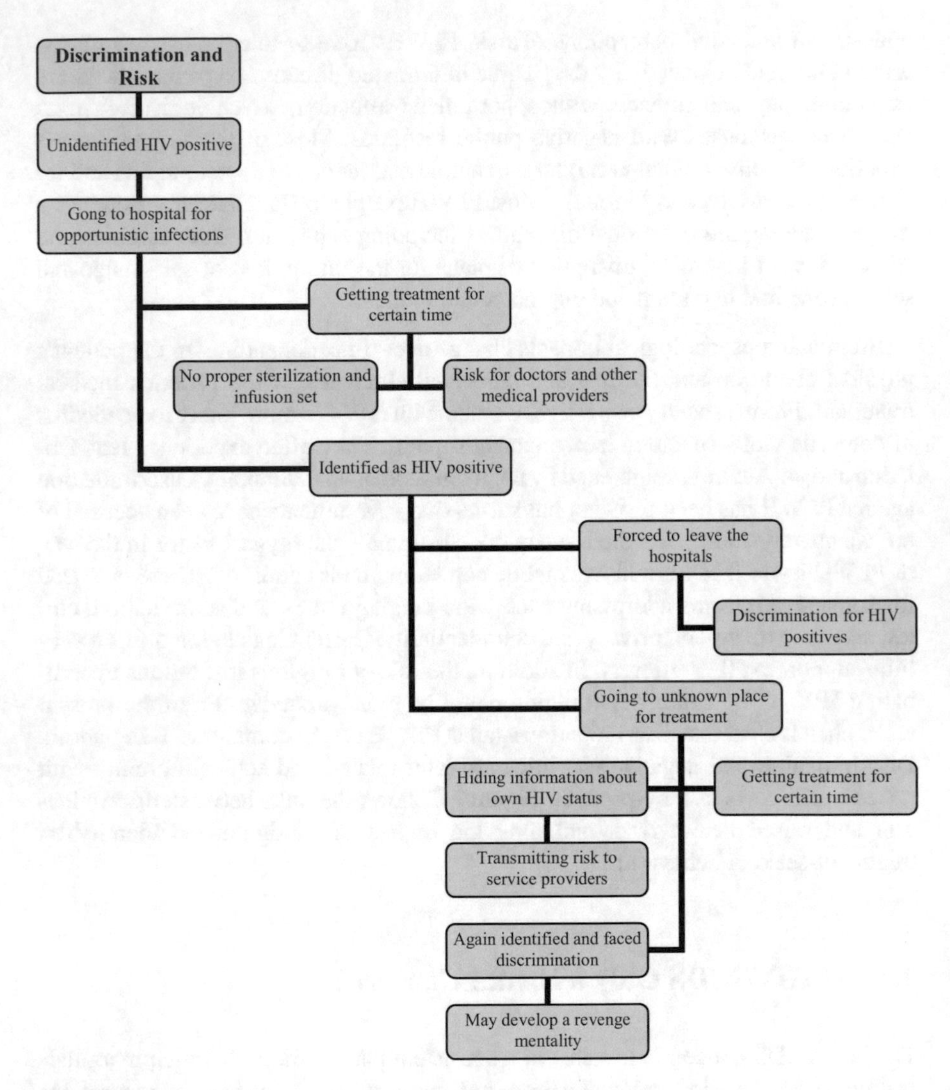

Fig. 7.2 Relationship between discrimination and health risk

On World AIDS day, 2007, the United Nations Secretary-General Ban Ki-moon called on governments around the world to allow universal access to HIV prevention and treatment and said that the stigma associated with the disease is the biggest barrier towards combating AIDS. In Bangladesh, article 27 and 28 (1) of the constitution state respectively that 'The state shall not discriminate against any citizens on the grounds of religion, race, cast, sex or place of birth' and 'All citizens are equal before law and are entitled to equal protection of law', but inequality before the law is clearly taking place and making already marginalized people more vulnerable.

7.4 Dilemmas in HIV Research in Bangladesh

There is a body of work on HIV/AIDS in Bangladesh but this work tends to adopt a quantitative and epidemiological approach. HIV/AIDS is therefore conceptualized as a bio-medical issue with risks which can be quantified and objectified. Rates of needle sharing and condom use, for example, are used as indicators for quantifying risk. The main objective of HIV awareness programmes in Bangladesh is to control the epidemic through prevention. There is a significant research gap which the present project has sought to address by using qualitative methods to explore, for example, the rationale for low use of condoms or high rates of drug use including needle sharing. That is not to negate the value of the bio-medical approach to understanding the spread of HIV infection. Rather, in order to understand the individual, structural, medical, cultural and socio-geographical contexts of HIV transmission and prevention in a conservative society such as in Bangladesh, both qualitative and quantitative approaches are necessary.

Information from in-depth interviews, focus group discussions, participant observations and round table discussions has elicited important findings in terms of understanding the 'lifeworlds' and health risks of HIV/AIDS in vulnerable people in two fieldwork sites. As a culturally sensitive issue, there is a large research gap present here. The Bangladeshi government is collecting some data through sero-surveillance for HIV but this is very limited in scope. As a result, policy planners face a scarcity of data about the health as well as social context of marginalized high-risk groups and bridging populations. Without core data on these groups, planning for more effective measures for solutions is clearly limited. This research specifically addresses the themes of health and social management with the help of an interest in the geographical settings. One of the prime outputs of this research is a methodology of investigation for the study of the complex health and social environment in Bangladesh.

7.5 Conclusion

Bangladesh has dealt with the causalities of many natural disasters, particularly devastating cyclones, for a long time. HIV/AIDS, a silent disaster, is still considered by many to be a low-level threat but the potential is there for losses from HIV/AIDS to impact on the nation's demographic, social and economic progress in future. AIDS is thought to be a disease of African or Indian people, whose sexual mores are different. But, despite being a Muslim country, Bangladesh has a long list of risk issues, including extramarital sex which could fuel the spread of HIV/AIDS. Bangladesh is said to have a low prevalence of HIV, but reported statistics are unlikely to be wholly reliable. The NASP has produced the 4th National Strategic Plan for HIV and AIDS response (2018–2022) that is comprehensive but the problem is with implementation due to resource constraints. Many NGO key officials believe that the government has received enough funding from different projects in the prevention of HIV but the problem is the channelling of funds. If the government can allocate resources to the right organizations at the right time, then there are opportunities to delay the epidemic. If the government's development partners and local civil society can make an intensive effort to work more closely, including involvement of marginalized communities themselves, positive results could be achieved.

Index

A. Paul, *HIV/AIDS in Bangladesh*, Global Perspectives on Health Geography,
https://doi.org/10.1007/978-3-030-57650-9

Printed in the United States
by Baker & Taylor Publisher Services